VITAL SIGNS

2003

Other Norton/Worldwatch Books

State of the World 1984 through *2003*
An annual report on progress toward a sustainable society

Vital Signs 1992 through *2002*
An annual report on the environmental trends that are shaping our future

Saving the Planet
Lester R. Brown
Christopher Flavin
Sandra Postel

How Much is Enough?
Alan Thein Durning

Last Oasis
Sandra Postel

Full House
Lester R. Brown
Hal Kane

Power Surge
Christopher Flavin
Nicholas Lenssen

Who Will Feed China?
Lester R. Brown

Tough Choices
Lester R. Brown

Fighting for Survival
Michael Renner

The Natural Wealth of Nations
David Malin Roodman

Life Out of Bounds
Chris Bright

Beyond Malthus
Lester R. Brown
Gary Gardner
Brian Halweil

Pillar of Sand
Sandra Postel

Vanishing Borders
Hilary French

VITAL SIGNS 2003

The Trends That Are Shaping Our Future

WORLDWATCH INSTITUTE

Michael Renner, *Project Director*
Molly O. Sheehan, *Associate Project Director*

Erik Assadourian
Arunima Dhar
Gary Gardner
Brian Halweil
Nicholas Lenssen
Lisa Mastny
Danielle Nierenberg
Radhika Sarin
Janet Sawin
David Taylor
Howard Youth

Linda Starke, *Editor*
Lyle Rosbotham, *Designer*

In Cooperation with the United Nations Environment Programme

W. W. Norton & Company
New York London

Printed in the United States of America
First Edition

The text of this book is composed in ITC Berkeley Oldstyle with the display set in Quadraat Sans.

Composition by the Worldwatch Institute; manufacturing by the Haddon Craftsmen, Inc.
Book design by Elizabeth Doherty and Lyle Rosbotham.

ISBN 0-393-32440-0 (pbk)

W. W. Norton & Company, Inc.
500 Fifth Avenue, New York, NY 10110
W. W. Norton & Company Ltd.
75/76 Wells Street, London W1T 3QT

1234567890

This book is printed on recycled paper.

Worldwatch Institute Board of Directors

Worldwatch Institute Staff

Contents

PART ONE: Key Indicators

PART TWO: Special Features

Acknowledgments

"Books are like imprisoned souls," wrote British author Samuel Butler, "till someone takes them down from a shelf and frees them." We owe a debt of gratitude to you, our readers, for freeing the information contained in *Vital Signs* in imaginative ways. From educators to activists, journalists to government officials, readers of this volume have written to tell us how they have brought *Vital Signs* to life—in the United States and Canada, many countries in Europe, Mexico, Argentina, Senegal, South Africa, Pakistan, Australia, Thailand, the Philippines, and Japan.

A psychology professor, for instance, was motivated by *Vital Signs* to design a graduate course on human behavior and the environment, while a clinical psychologist uses the book in educational presentations to physicians, linking the health of the planet to human health. In Australia, *Vital Signs* is a "must have" among those who train advocates and educators for a national environmental organization. An environmental advisor to a major oil company wrote to let us know that he depends on the book to grasp "the larger picture." Print, radio, and television journalists in Argentina, the United Kingdom, South Africa, Senegal, and Pakistan rely on *Vital Signs* as a reference and research tool.

Well before the book is published, we depend on a wide range of experts who donate their time to critique drafts or provide key data. For their honest reviews, useful comments, and other help this year, we thank Sarah Joy Albrecht, Cornia Andrea, Alyssa Becker, Stan Bernstein, Joseph Chamie, Claudine Chapeau, Roberta Cohen, Nigel J. Collar, Colin Couchman, Seth Dunn, Emilio Escudero, Majid Ezzati, Katya Fay, Michael Flynn, David Fridley, Christian Friesendorf, Satoshi Fujino, Don Hinrichsen, Alvin Hutchinson, Ragupathy Kannan, Tim Kelly, Philippe LeBillon, Wilfrid Legg, Wayne Lo, Birger Madsen, Tim McGirk, Dan McMeekin, Corin Millais, Robert Nicholls, David Roodman, Martha Rosen, Michael Ross, Hiram Ruiz, Henry Saffer, Payal Sampat, Wolfgang Schreiber, Courtney Ann Shaw, Vladimir Slivyak, Adam Smith, Alison J. Stattersfield, Victor Strasburger, Shiyun Sung, Stefanie Teggemann, Arnella Trent, Mathis Wackernagel, John Washburn, Timothy Whorf, Angelika Wirtz, and Jennifer Woofer.

This is the third year in which the United Nations Environment Programme (UNEP) has cooperated with Worldwatch to produce *Vital Signs*. We have greatly enjoyed working with UNEP's Marion Cheatle, who helped us overcome the barrier of distance between Worldwatch's office in Washington and UNEP's headquarters in Nairobi for this team effort. And we thank Munyaradzi Chenje and So-Young Lee for their careful reviews. Our colleague Hilary French acted as a bridge between the two organizations.

After outside experts and UNEP colleagues review the text and graphs, we send the completed *Vital Signs* manuscript to our longtime publisher, W.W. Norton & Company, in

New York. We rely on the considerable talents of Norton's Amy Cherry, Lucinda Bartley, Bill Rusin, Leo Wiegman, and Andrew Marasia to speed the manuscript through production and into bookstores and classrooms across the United States.

We are also indebted to our international partners who publish *Vital Signs* in many languages outside the United States. For their tireless efforts on last year's *Vital Signs*, we thank Soki Oda of Worldwatch Japan, Anna Bruno Ventre of Edizioni Ambiente in Italy, Gianfranco Bologna of WWF Italy, Sang Baek Lee and Jung Yu Jin of the Korean Federation for Environmental Movement, Lluis Garcia Petit and Sergi Rovira at Centro UNESCO de Catalunya in Spain, Marisa Mercado at Fundación Hogar del Empleado in Spain, Eduardo Athayde of UMA–Universidade Livre da Mata Atlantica in Brazil, Eilon Schwartz of the Heschel Center for Environmental Learning and Leadership in Israel, and Hamid Taravati in Iran.

Many contributions to this annual book stem from Worldwatch's general research program, which is backed by a roster of philanthropic organizations. We thank the following foundations for their generous support over the last year: the Aria Foundation, the Richard & Rhoda Goldman Fund, The George Gund Foundation, The William and Flora Hewlett Foundation, The Frances Lear Foundation, The John D. and Catherine T. MacArthur Foundation, the Merck Family Fund, the Curtis and Edith Munson Foundation, Nalith, Inc., the NIB Foundation, The Overbrook Foundation, The David and Lucile Packard Foundation, The Shared Earth Foundation, The Shenandoah Foundation, Turner Foundation, Inc., the Wallace Global Fund, the Weeden Foundation, and The Winslow Foundation.

We also greatly appreciate the support we receive from thousands of Friends of Worldwatch, and give special thanks to our Council of Sponsors: Adam and Rachel Albright, Tom and Cathy Crain, John and Laurie McBride and Kate McBride Puckett, and Robert Wallace and Raisa Scriabine. We were saddened in October 2002 by the death of Bob Wallace, a longtime friend of Worldwatch who inspired countless people through his work to build a better world.

In recent years Worldwatch's Board of Directors has stepped up its commitment to strengthen the Institute. We thank this exceptional group of people for their inspired leadership over the last year in guiding the Institute through a challenging period of political and economic uncertainty in the world.

Bolstered by this tremendous network of colleagues, donors, and friends, Worldwatch staffers bring a high level of enthusiasm to *Vital Signs*. Elizabeth Nolan, in charge of business development, coordinates our efforts with Norton and our international partners. Our development team of Adrianne Greenlees, Kevin Parker, Mary Redfern, and Cyndi Cramer continues to expand Worldwatch's broad base of support, while our communications team—Leanne Mitchell, Susan Finkelpearl, and

Susanne Martikke—brings *Vital Signs* to ever larger audiences. Integral to these efforts are Barbara Fallin, our Director of Administration, in-house Internet and technology experts Patrick Settle and Steve Conklin, and Joseph Gravely in our mailroom.

To cover the diverse topics in *Vital Signs*, we draw on the work of many researchers, from former staffers to recent hires. Worldwatch alumni Howard Youth and Nick Lenssen served as authors this year, while Seth Dunn, David Roodman, and Payal Sampat provided guidance on early drafts. Research librarian Lori Brown helped authors find crucial reports and data sets. Senior Researcher Chris Bright provided particularly helpful reviews, as did Senior Researcher Janet Sawin, who also spent hours calculating carbon emissions in addition to writing her own pieces. The contributions this year of Staff Researchers Erik Assadourian and Radhika Sarin were especially useful in uniting the book around the themes of poverty and inequality. The research staff also relied heavily on interns-turned-authors Arunima Dhar and David Taylor, and on Anand Rao, who joined the intern staff just in time to jump into assembling data for *Vital Signs*.

Each year, the responsibility of transforming a multitude of contributions into a coherent book—under deadline pressure—falls to our editor and our art director. As always, independent editor Linda Starke transformed drafts into polished prose with unparalleled speed. This year marked Art Director Lyle Rosbotham's first foray into *Vital Signs*, and he added his own personal touch to many design elements. Several of the photos Lyle chose for the opening pages of sections are from Photoshare, the online database of the Media/Materials Clearinghouse at the Johns Hopkins University Population Information Program at <www.jhuccp.org/mmc>.

We hope that our readers will continue to bring this book to life. The data used to prepare all of the figures in this book are on our CD-ROM, *Signposts 2003* (see p. 155). Please let us know if you have ideas of trends to be covered in future editions. You can reach us by e-mail (worldwatch@worldwatch.org), fax (202-296-7365), or regular mail.

Michael Renner and Molly O. Sheehan
March 2003

Worldwatch Institute
1776 Massachusetts Ave., N.W.
Washington, DC 20036

Preface

The past year was marked by frequent, graphic reminders of the human costs of environmental disruption. In May and June, South Asia was hit with a dangerous heat wave, and temperatures as high as 50 degrees Celsius led to the deaths of more than 1,200 Indians in a single week. A month later, heavy monsoon rains led to mudslides and flooding in parts of India, killing another 300 people and disrupting the lives of more than 10 million. Other areas of India were cruelly neglected by the year's monsoon, causing the most severe drought in 15 years.

Statistics published in *Vital Signs 2003* show that economic losses from weather-related disasters worldwide totaled $53 billion in 2002. Although this figure is short of the record $100 billion in losses in 1999, it suggests that the unprecedented losses from weather-related disasters that hit the world in the 1990s are continuing into this new decade.

Weather-related economic losses were highest in industrial countries in 2002—August floods in central Europe cost the region $18.5 billion—but the human toll was far higher in developing countries. Heavy rains in Kenya displaced 150,000 people, while 800,000 people in northern China suffered from a severe drought.

These statistics reflect the intersection of two powerful global forces—growing environmental degradation and stubbornly high levels of poverty. Poor people living in precarious conditions are the most vulnerable to storms and floods, which are made worse by deforestation, soil erosion, and climate change.

The human tragedies behind the statistics are a compelling reminder that environmental and social progress are not luxuries that can be set aside when the world is experiencing economic and political problems. Rather, they are central to human well-being. Unless the world can make better headway in improving environmental and social health in the years ahead, the toll of weather-related disasters will continue to rise.

Vital Signs 2003 reports many ways in which the benefits of a growing global economy are still not reaching billions of people. These are reflected most starkly in global health statistics. For example, infectious diseases kill over 14 million people each year, most of them in developing countries.

Environmental ills are also amply documented in this year's *Vital Signs*: in 2002, the global average temperature reached the second highest level ever recorded, sustaining a warming trend that climate scientists believe is linked to the atmospheric buildup of carbon dioxide from fossil fuel burning. And ornithologists are documenting the decline of birds: some 12 percent of the world's 9,800 bird species are threatened with extinction in this century, largely because human activities are destroying their habitat.

Just as poverty and environmental decline come together to exacerbate the effects of natural disasters, the world community must address these and related issues with a combined strategy if it is to have any chance of successfully addressing each of them. Unless natural systems are stabilized, human welfare will be undermined. And unless excessive consumption is effectively addressed and the needs of the poor are met, both these pressures will continue to strain the health of the world's ecosystems.

Two recent global agreements provide a valuable framework for addressing poverty and environmental decline in tandem. The Millennium Development Goals (MDGs) adopted by the United Nations General Assembly in 2000 have helped galvanize and focus international efforts on such objectives as eradicating poverty and hunger, achieving universal primary education, promoting gender equality, ensuring environmental sustainability, and developing a global partnership for development.

The Plan of Implementation that was adopted by governments at the World Summit on Sustainable Development in Johannesburg in 2002 reaffirmed the MDGs. It also highlighted that eradicating poverty, changing unsustainable patterns of production and consumption, and protecting and managing the natural resource base are overarching objectives of and essential requirements for sustainable development. The Plan included several new targets, including restoring fisheries, stabilizing biological diversity, and meeting the sanitation needs of a half-billion people.

None of these goals will be easy to achieve. And only through strengthened commitments by governments, international institutions, and civil society is there any chance of success. In 2002, U.N. Secretary-General Kofi Annan summed up the challenge in a report to the General Assembly on implementation of the MDGs, stating that despite some signs of progress, "the world community has a long way to go towards fulfilling the (Millennium) Declaration's goals." These words echo his statement during the Millennium Summit, when he said: "There is much to be grateful for. There are also many things to deplore, and to correct."

Among the trends covered in *Vital Signs 2003*, we believe that you will find much to be grateful for and also much to deplore. We sincerely hope that this book will help the world community correct its course.

Christopher Flavin
President
Worldwatch Institute

Klaus Töpfer
Executive Director
United Nations Environment Programme

VITAL SIGNS

2003

TECHNICAL NOTE

Units of measure throughout this book are metric unless common usage dictates otherwise. Historical population data used in per capita calculations are from the Center for International Research at the U.S. Bureau of the Census. Historical data series in *Vital Signs* are updated each year, incorporating any revisions by originating organizations.

Unless otherwise noted, references to regions or groupings of countries follow definitions of the Statistics Division of the U.N. Department of Economic and Social Affairs.

Data expressed in U.S. dollars have for the most part been deflated to 2001 terms. In some cases, the original data source provided the numbers in deflated terms or supplied an appropriate deflator, as with gross world product data. Where this did not happen, the U.S. implicit gross national product (GNP) deflator from the U.S. Department of Commerce was used to represent price trends in real terms.

Overview

Poverty and Inequality Block Progress

Michael Renner and Molly O. Sheehan

Environmental protection and human well-being are critical challenges. How do we protect Earth's fragile ecosystems without denying billions of people a chance for a better life? How do we improve the human condition without wrecking the delicate balance that sustains all life on the planet? International conferences—most recently, the World Summit on Sustainable Development in 2002—have brought attention to these questions, scholars are debating underlying theories and strategies, and grassroots activists are putting forth their own proposals for change.

These twin goals cannot be achieved as long as humanity remains divided into the extremes of rich and poor. A sizable minority enjoys plentiful food, seemingly unlimited mobility, access to cutting-edge technology, and other amenities of modern life. Others are struggling to make ends meet, while the large majority of people has scant opportunity to look past the worries of daily survival. Two different types of environmental destruction result: the wealthy impose the heaviest toll on the planet by dint of their materials-intensive, pollution-laden lifestyles, whereas the poor generally live with some of the worst local environmental conditions, eking out a meager living only by taxing their croplands, forests, and water resources to the limits.

Globalization—increased trade, investment, travel, and other border-transcending changes—has deepened these disparities. It has been an engine of unrivaled economic opportunity for some and a source of increasing pressure and anxiety for many more. The world economy has grown sevenfold since 1950. (See pages 44–45.) But instead of this rising tide lifting all boats, the smaller vessels are in danger of sinking or have already capsized.

Poverty is first of all a lack of sufficient income, with more than 2 billion people worldwide struggling to survive on a few dollars a day or less. Hunger is a widespread phenomenon on this planet—but people go hungry not because of a scarcity of food, but because they are too poor to buy enough. Most of the hungry live in South Asia and sub-Saharan Africa. A substantial share of world grain supplies is sold as food not for hungry people but for livestock. And then most meat is eaten in affluent parts of the world, with people in industrial countries consuming about three times as much meat as people living in developing countries. (See pages 28–31.)

As this example suggests, in a world that offers a portion of humanity unprecedented wealth and opportunity, poverty cannot be understood without also looking at inequality. Poverty and inequality have many dimensions, ranging from the subtle to the extreme, from the obvious to the surprising, and from the fleeting to the deeply structural. And it is not enough to speak only of rich and poor countries. Deep disparities are also found within individual countries.

As many of the contributions to this book illustrate, poverty and inequality manifest themselves in highly unequal educational opportunities, heightened vulnerability to preventable and curable diseases, and a gaping digital divide. For the poor, this translates into underfunded social programs due to crushing foreign debt burdens, greater exposure to armed conflict and human rights violations, and heightened susceptibility to natural disasters. In fact, the consumption choices of the rich often unleash forces—pollution, waste, resource depletion, and environmental damage—that hit the poor hardest, even though they are far less able to protect themselves against resource depletion, droughts, and other extreme weather events.

A Growing Economic Divide

Between 1960 and 1995, the disparity in per capita income between the world's 20 richest and 20 poorest nations more than doubled, from 18 to 1 to 37 to 1. But because inequality also rose within most countries, the gap between those at the top and those at the bottom of the economic pyramid is even more pronounced. Inequality has risen not only in poor countries and in those struggling in transition from communist to capitalist economic systems, but also in many industrial nations. (See Table 1 and pages 88–89.)

One measure of growing disparities within wealthy nations is the widening gap between the compensation of corporate chief executive officers (CEOs) and the pay of employees. That differential grew more than fivefold during the 1990s in the United States, where it is by far the most pronounced. CEOs there made 350 times as much as the average factory worker in 2001, and sometimes were awarded lavish stock options even as layoffs were announced. This "pay gap" is as least 10 times as large in the United States as in other industrial nations, but as transborder economic connections multiply and comparable practices are adopted elsewhere, similar outcomes are more likely in other countries. (See pages 90–91.)

Many corporate and government leaders insist that stimulating additional economic growth is key to achieving society's goals. It is true that even small increases in the income of the poor can translate into dramatic health and education benefits for them. But evidence and common sense suggest that growth under conditions of high inequality brings few benefits to the "have-nots," does little to reduce poverty, and may even constrain future economic growth. (See pages 88–89.) In addition, corruption saps economic development and skews public investment away from the priority areas of education and health that are most likely to reduce poverty. (See pages 114–15.)

In a variety of ways, the world economy is rigged against the interests of the poor. Farm subsidies of more than $300 billion per year, for example, allow food crops exported by farmers in industrial countries to be sold at prices 20–50 percent below the cost of production, undermining farmers in developing nations. (See pages 96–97.)

In Mexico, Peru, and Colombia, farmers are turning to drug crops like opium, coca, or cannabis because their food crops cannot compete with cheaper, mass-produced imports. In

Table 1: Income Inequality in Selected Countries, 1990s

	Share of Income		
Country	Richest 20 Percent	Poorest 20 Percent	Ratio
	(percent)		
Japan	35.7	10.6	3.4 to 1
Denmark	34.5	9.6	3.6 to 1
Egypt	39.0	9.8	4.0 to 1
Indonesia	41.1	9.0	4.6 to 1
France	40.2	7.2	5.6 to 1
Yemen	41.2	7.4	5.6 to 1
India	46.1	8.1	5.7 to 1
United Kingdom	43.2	6.1	7.1 to 1
China	46.6	5.9	7.9 to 1
United States	46.4	5.2	8.9 to 1
Russia	53.7	4.4	12.2 to 1
Nigeria	55.7	4.4	12.7 to 1
Mexico	57.4	3.5	16.4 to 1
Zambia	56.6	3.3	17.2 to 1
Brazil	64.1	2.2	29.1 to 1

Source: World Bank. Data are for most recent year available.

Afghanistan, warfare destroyed much of the legal crop base. Drug crops are lucrative because of high demand in wealthy nations. Even so, the largest share of profits goes not to farming communities but to those controlling the retail trade. (See pages 98–99.)

On the flip side, developing-country exports, particularly of agriculture and textiles, typically face trade barriers twice as high as industrial-country exports. Unequal trade relations are one reason why developing and former Eastern bloc nations are saddled with huge foreign debts that totaled $2.44 trillion in 2001. In diverting scarce resources to debt repayment, many governments fail to meet the basic health and education needs of their populations. (See pages 46–47.)

Diseases of Poverty and Wealth

Less income all too often translates into poorer health, greater mortality, and shorter life expectancy for the world's "have-nots." The infant mortality rate in low-income countries is 13 times the rate in high-income countries. (See pages 88–89.) Maternal mortality also reveals a stark divide between rich and poor: some 99 percent of all pregnancy-related deaths occur in developing countries, where women face a lifetime risk of maternal death that is 40 times greater than that of women in industrial nations. (See Figure 1.) In the 49 least developed countries in the world, women are at least 150 times as likely to die due to pregnancy or childbirth. (See pages 106–07.)

Many people in the developing world—particularly those earning less than $2 a day—are tormented by an array of infectious diseases. These threats to public health are a product of pervasive poverty: the lack of clean water, sanitation, affordable medicines, and nutritious food. With just 36 percent of the global population, Africa and Southeast Asia account for 75 percent of deaths from such diseases. (See pages 108–09.)

AIDS is one of the infectious diseases that disproportionately kill poor people. Since 1996, antiretroviral drugs have dramatically reduced AIDS deaths in high-income countries. But even with a sharp drop in the price of a year's worth of these drugs (to $350 per person), only 4 percent of people living with AIDS in low- and middle-income nations can afford them. (See pages 68–69.)

Economic inequities have hastened the rapid spread of HIV among women, who for the first time account for half of those living with the virus. While biological differences make women more susceptible than men to HIV, the economic disadvantages that women face in most societies heighten their vulnerability, as women who are economically dependent on husbands or sexual partners often have little control over sex and condom use.

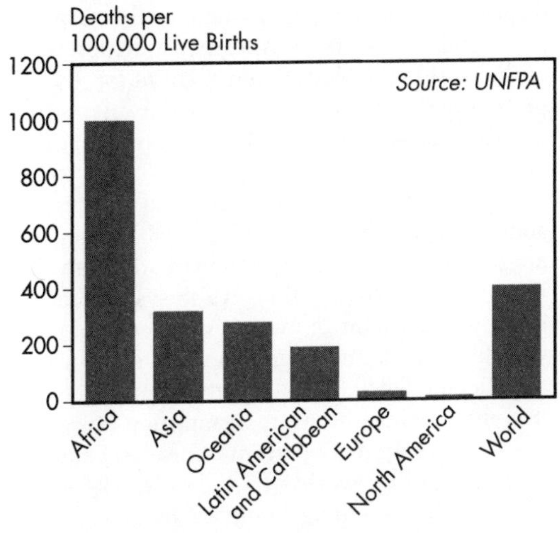

Figure 1: Maternal Mortality Rates, by Region, 1995

Not only do the poor account for the bulk of AIDS victims, but the disease also worsens poverty, as those who succumb to it are often in the prime of their lives: parents and wage earners. Without AIDS, the number of children worldwide who are orphans would be declining; instead, their ranks are growing. By 2010, some 25 million children will have lost one or both parents to this disease. (See pages 110–11.)

Other risks of poverty include dietary deficiencies, which increase susceptibility to infectious diseases and are responsible for up to 6 million deaths annually. Without enough money to buy food, some 815 million people worldwide are chronically hungry. (See pages 28–29.) Lack of clean water or sanitation kills 1.7 million people each year—90 percent of them children—and indoor smoke from heating and cooking fuels causes 1.6 million deaths. (See pages 108–09.)

Even as the "underconsumption" of clean water, food, education, or medicines accounts for up to 23 percent of deaths worldwide, the World Health Organization estimates that the "overconsumption" of food, tobacco, alcohol, and drugs accounts for up to 46 percent of mortality. Wealthier parts of the world have a disproportionate amount of such deaths. Europeans and Americans constitute just 28 percent of the world's population, but account for 42 percent of deaths from cardiovascular diseases and cancers—diseases of affluence. (See pages 108–09.)

Diseases related to unhealthy food choices, addictive substances, and sedentary lifestyles are a growing threat as well in many developing countries, where public health systems are already overwhelmed. Even in poor nations with high death rates, where poor sanitation and dietary deficiencies account for up to 42 percent of all deaths, overconsumption now accounts for up to 27 percent of deaths. Some 82 percent of the world's 1.1 billion smokers now live in developing countries; by 2030, experts predict that 7 million out of 10 million smoking-related deaths will occur in the developing world. (See pages 70–71.)

Poverty and Armed Conflict

Virtually all of the world's current armed conflicts take place in the developing world. Wars deepen the poverty and deprivation of civilian communities caught directly in the fighting, uprooted by violence, or hit by the repercussions—ruined economies, destroyed public infrastructures, and damaged public health systems. (See pages 74-75.)

Sadly, this is particularly the case where conflict ignites over resource wealth—commodities like oil, minerals, and gemstones. Governments have used resource revenues to purchase arms rather than fund social programs. And warlords and other predatory armed groups are resorting to extreme violence in order to gain control of lucrative resources. (See pages 114–15 and 120–21.)

Due to widespread warfare and political repression, developing countries generated 86 percent of the world's refugees during the past decade. Refugees, internally displaced people, and environmental refugees (those uprooted by natural disasters or severe environmental degradation) worldwide now number more than 60 million. (See pages 102–03.)

The international community continues to make halting efforts to address conflicts and human rights violations, and to bring about stable conditions that allow war-torn areas to emerge from destruction and poverty. The United Nations and other international organizations conducted close to 50 peacekeeping operations in 2002, involving some 110,000 soldiers and police. (See pages 76–77.) And in 2002, the International Criminal Court opened its doors, advancing the international community's efforts to deter genocide, war crimes, and other crimes against humanity. Unfortunately, this effort is being opposed by a small number of countries, including the United States and China. (See pages 116–17.)

The renewed growth in world military expenditure has troubling implications for many of the world's unmet needs, including health services, education, poverty alleviation, and environmental protection. Some of the wealthier nations, particularly the United States, are once more giving priority to building military muscle over foreign aid and debt forgiveness. Low-income countries themselves account for only a tiny share of global military budgets—about 7 percent. Even so, this is more than double their share of the world's gross economic product, and military spending is a heavy burden for these impoverished nations. For countries such as Eritrea,

Burundi, and Pakistan, military spending equals or surpasses combined public expenditures for health and education. (See pages 118–19.)

The Inequities of Climate Change

Poor nations and communities are disproportionately vulnerable to the reverberations of climate change. Global average temperature climbed to 14.52 degrees Celsius in 2002, supplanting 2001 as the second hottest year since recordkeeping began in the late 1800s. The nine warmest years have all occurred since 1990. Scientists expect that the temperature record set in 1998 will be broken with a new high in 2003. (See pages 40–41.)

Scientists predict that higher global temperatures will translate into a greater number of extreme weather events. The frequency and duration of El Niño patterns, which disrupt weather worldwide, have increased in recent years. In 2002, El Niño–related disasters helped spur a 93-percent increase in insured losses over the previous year. (See pages 92–93.)

Weather-related disasters take the largest human toll in developing nations. In 2002, an erratic summer monsoon pounded eastern India, Nepal, and Bangladesh with rain and floods, while other parts of India suffered the worst drought in years, as an unprecedented heat wave caused the highest one-week death toll in the nation's history. At the same time, record-breaking rains, floods, and landslides injured more than 80,000 people in China.

A warming world also means rising sea levels, which threaten the very existence of small islands states. The Intergovernmental Panel on Climate Change estimates that global sea level will accelerate in the coming century. In the worst-case scenario, nations such as the Marshall Islands, Tuvalu, and Kiribati could disappear entirely. (See pages 84–85.)

While the burden of climate-related disasters falls most heavily on developing nations least able to adapt, the burden of responsibility lies with the industrial nations that have, throughout history, contributed the bulk of carbon emissions through fossil fuel burning. With less than 5 percent of the world's population, the United States is the single largest consumer of oil, coal, and natural gas. (See pages 34–35.)

The United States is accordingly the largest contributor to climate change, producing 24 percent of the world's carbon emissions from fossil fuel burning. The U.S. passenger car fleet alone, accounting for one quarter of the world total, produces as much carbon as the entire Japanese economy. (See pages 56–57.) Per person, U.S. carbon emissions are roughly double that of other major industrial nations, and 17 times that of India. (See Figure 2 and pages 40–41.)

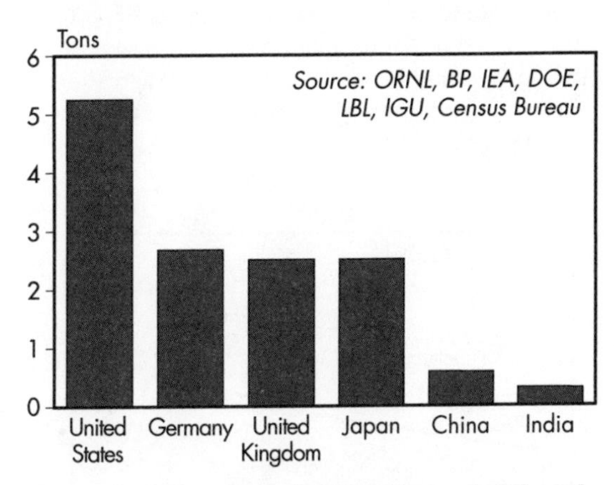

Figure 2: Carbon Emissions Per Person in Selected Countries, 2002

China, home to one fifth of the world's people, ranks a distant second to the United States in total emissions, with just 12 percent of the world's total. (See pages 40–41.)

Communication Gaps

As with energy use and climate change, the benefits and costs of electronic communications technologies are unevenly distributed. By linking farmers to market information, workers to customers, patients to doctors, and students to teachers, the Internet can aid economic development. Internet use, however, is still concentrated in industrial nations, where there are 41 Internet users per 100 people compared with just 2.3 per 100 in developing nations. This 17 to 1 ratio, while huge, is down from 40 to 1 in 1995. (See pages 60–61.)

The enormous gap in phone access between rich and poor nations is also shrinking, largely thanks to new mobile phone operations, which are cheaper to build than conventional, fixed-line systems. In a few years, mobiles have come to dominate Africa: in 1999, Uganda became the first African country to have more mobile than fixed-line customers; some 30 other nations quickly followed, so that mobiles now outnumber fixed lines in Africa at a higher ratio than on any other continent. Still, industrial nations have more phones than people—121 phone links per 100 people, whereas the poorest nations have barely more than 1 phone connection per 100 people. (See Figure 3 and pages 60–61.)

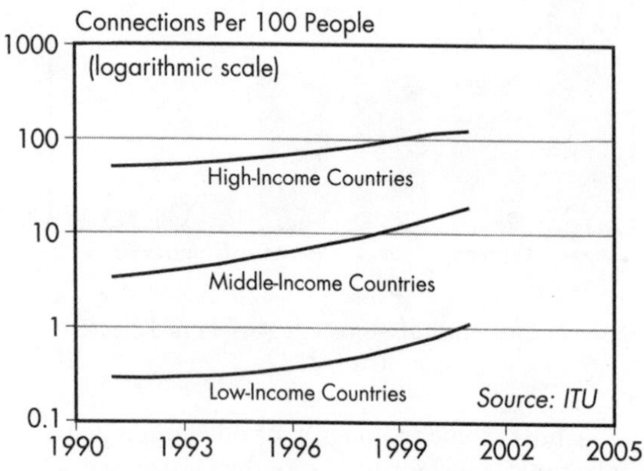

Figure 3: Phone Connections (Cellular and Fixed-Line) by Income Level of Country, 1991–2001

While computers and phones benefit people in many ways, their assembly requires large amounts of water, energy, and toxic materials, which can harm drinking water supplies and people, especially those who make the devices and those who recycle them. Both the manufacture and disposal of electronic technologies are increasingly concentrated in Asia. Most semiconductor makers, including those in North America and Europe, outsource production to a few large foundries in Singapore and Taiwan, while nations such as China and India receive large amounts of "e-waste" for recycling. (See pages 62–63.)

Standing Apart: The United States and Africa

Anyone searching for the most pronounced examples of rich and poor on our planet need look no further than the United States and much of the African continent. These regions are at the opposite ends of a spectrum that separates opportunity and vulnerability and that illustrates extravagant resource use and the pervasive lack of even basic amenities. Many of the contributions to this year's *Vital Signs* highlight this enormous contrast and contradiction in the human experience.

The United States, with about 5 percent of world population, accounts for 22 percent of the global economic product, 25 percent of the passenger cars, and more than half of what the world spends on advertising. (See pages 44–45, 48–49, and 56–57.)

As the world's single largest contributor to carbon emissions, the United States is doing more than any other nation to warm the global atmosphere. It is therefore striking that the United States has abandoned the Kyoto Protocol to combat climate change while most of the world is moving forward to adopt it. By the end of January 2003, more than 100 nations had ratified the protocol, including the 15 nations of the European Union, Japan, and Canada. (See pages 40–41.) At the same time, last year the U.S. administration announced plans for tax credits that would promote the most fuel-inefficient passenger vehicles. (See pages 56–57.)

Africa, in stark and dismal contrast, is home to 70 percent of the world's HIV-positive people. Sub-Saharan Africa is being dragged down by the AIDS epidemic, which is now the leading cause of death there. In

2002, average life expectancy in 16 African nations was at least 10 years lower than it would have been without AIDS. The food crisis in Africa has been exacerbated by the loss of agricultural workers to AIDS and by savings being spent on medicines and funerals. (See pages 68–69.)

Sub-Saharan Africa also leads the world in pregnancy-related deaths. No other region comes close to Africa's 1,000 maternal deaths per 100,000 live births. In fact, while gains have been made in many world regions, delivery care has actually worsened in some African nations, where fertility levels remain high. (See pages 106–107.)

Resource wars are further devastating many African nations ravaged by AIDS and buckling under health crises. (See Figure 4.) These include violent conflicts fueled by diamond and oil money in Angola, where nearly half of all children are underweight and almost 30 percent die before the age of six. Violent struggles linked to profitable natural resources have also plagued Sierra Leone, Liberia, Sudan, and the Democratic Republic of the Congo. (See pages 120–21.)

Bridging the Divides

At the root of sustainable development—meeting the needs of all today without endangering the prospects of future generations—is a more equitable distribution of resources and opportunities. Several indicators in this edition of *Vital Signs* highlight areas where some form of progress toward this goal can be seen.

While poverty and hunger persist, the number of hungry people worldwide has actually declined nearly 15 percent from 1970's figure of 956 million. (See pages 28–29.) And in recent years, activists advocating debt relief for the world's poorest countries have raised global awareness of poverty and helped spur the development of new debt forgiveness plans administered by the World Bank. (See pages 46–47.)

Since 1970, a slowdown in population growth and lower birth rates helped to improve development prospects in Brazil, Mexico, and

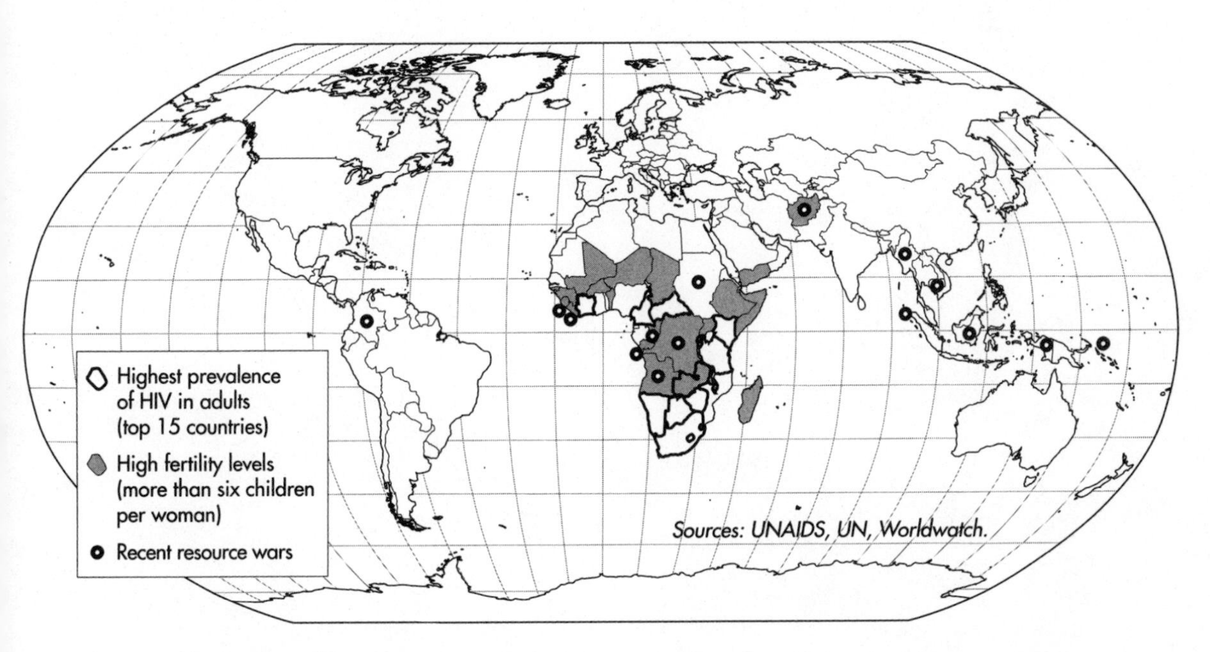

Figure 4: Countries Under Stress from HIV Epidemic, Population Growth, and Resource Wars

several East Asian countries. With better access to contraceptives, people had fewer children and more women could work outside the home. This opened a demographic window of opportunity for development, as a large group of working-age people were supporting a relatively smaller number of dependents. (See pages 66–67.)

New types of industry are beginning to create jobs with less pollution and waste generation. Wind turbines, for example, provide electrical power to a growing number of people. Global wind capacity has tripled since 1998, and remains the fastest-growing power source. Wind energy is now concentrated in Europe, but as costs fall, India and China are becoming larger users. (See pages 38–39.) Another key global industry, tourism, began to recover from recent setbacks. As Asia becomes a prime tourist destination, ecotourism holds promise for protecting the habitat of birds and other wildlife while providing needed jobs. (See pages 50–53 and 82–83.)

Greater investments in health, education, and innovative industries are needed to tap the human potential that is now bound by poverty. This could be accomplished with a reordering of priorities. The estimated funds required to eliminate starvation and illiteracy, to grant reproductive health care to all women, and to provide clean drinking water worldwide are far less than the money spent each year on military purposes. (See pages 118–19.) And in recent years, governments and international agencies have started to acknowledge the corrosive effect of corruption on development, which is skewing spending away from health and education and toward arms buildup and large public works. (See pages 114–15.)

Just as poverty has many dimensions, the struggle to overcome divisive disparities is a multifaceted undertaking, requiring technical innovation, social rejuvenation, and political change. While the complexity of the challenge may invite despair and inaction, it also offers many levers and opportunities for change.

Part One

KEY INDICATORS

Food Trends

© Digital Vision

Grain Production Drops

Meat Production and Consumption Grow

Grain Production Drops

Brian Halweil

In 2002, global grain production declined for the third time in four years, due mainly to drought in North America and Australia.[1] At 1,833 million tons, the harvest was 3 percent lower than the previous year's and was the smallest crop since 1995.[2] (See Figure 1.)

World grain production has more than doubled since 1961, mainly due to farmers harvesting more grain from each hectare, since farmers are planting grain on only slightly more land today—671 million hectares in 2002 compared with 648 million in 1967.[3] The average harvest of grain from a given hectare has more than doubled worldwide, from 1.24 tons in 1961 to 2.82 tons in 2002.[4]

LINKS pp. 30, 96

Still, production of the three major cereal crops declined in 2002. Global wheat production dropped to 562 million tons, down 3 percent; production of corn stood at 598 million tons, nearly 2 percent lower; and rice production, at 391 million tons, was 2 percent below output in 2001.[5] These three grains account for 85 percent of the world's grain harvest.[6]

Global grain production is concentrated geographically. China, India, and the United States alone account for 46 percent of global production. Europe, including the former Soviet states, grows another 21 percent.[7]

Of the world's major grain-producing regions, production across Asia was up, in Europe it was stagnant, and in the North American wheat and corn belts the harvest suffered from drought and high summer temperatures.[8] A weak and irregular monsoon reduced rice harvests in India and, to a lesser degree, in China—the world's second largest and top rice producers, respectively. The wheat harvest declined slightly in China (the world's largest producer) but was up in India (the second largest producer).[9]

In the United States, responsible for at least one third of the global corn harvest, a severe drought across the middle of the country cut production by 8 percent.[10] And the U.S. wheat harvest was down by 14 percent—also due to drought.[11] In Australia, severe drought reduced the grain harvest by almost 40 percent.[12]

Despite a shift toward more meat eating and greater dietary diversity around the world, people still primarily eat foods made from grain. On average, they get about 48 percent of their calories from grains, a share that has declined just slightly from 50 percent over the last four decades.[13] Grains, particularly corn, also form the primary feedstock for industrial livestock production.

Global grain production per person dropped to 294 kilograms in 2002, the lowest level since 1970.[14] (See Figure 2.) But output per person varies dramatically by region. For instance, it stands at roughly 1,046 kilograms in North America, most of which is fed to livestock, compared with 316 kilograms in China and just 120 kilograms in sub-Saharan Africa.[15]

While the downward slide in the global output per person could eventually prove problematic for food supplies, the focus on this number can be misleading, since people are hungry primarily because they are too poor to purchase food, not because of an outright scarcity of food. The U.N. Food and Agriculture Organization estimates that there are at least 815 million chronically hungry people in the world, a modest decline from the 956 million estimated in 1970.[16]

Most of these hungry people are concentrated in India and Asia, although the most acute increase in hunger in 2002 was in sub-Saharan Africa, where roughly 40 million people are in immediate need of food aid.[17] Following two consecutive years of poor grain harvests—exacerbated by drought, civil conflict, and HIV/AIDS—grain and flour prices in the region have increased beyond the reach of much of the population, and both imports and international relief have been insufficient to stem the rise.[18]

Global grain production exceeded consumption between 1996 and 1998, but the harvest has slipped below demand for the last four years, pushing down the stocks of grain held in private and government stores.[19] World cereal stocks fell sharply, to some 466 million tons, by the end of 2002—nearly a 20-percent reduction in just one year and the lowest level in 40 years of recordkeeping.[20] The ratio of grain stocks to annual use also hit an all-time low.[21] (See Figure 3.)

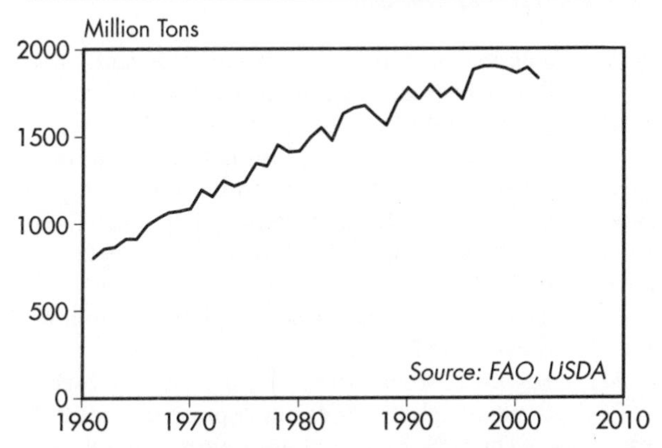

Figure 1: World Grain Production, 1961–2002

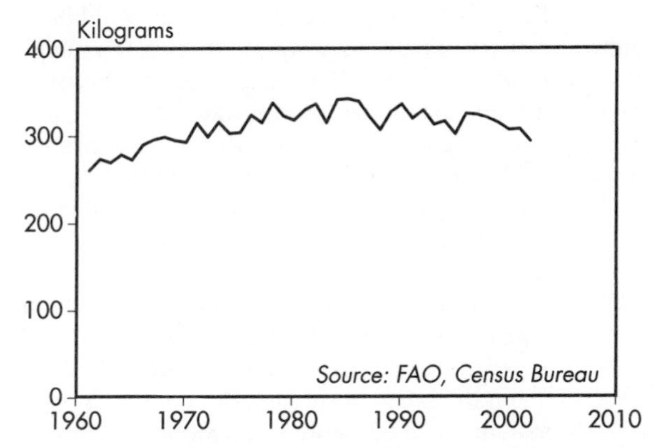

Figure 2: World Grain Production Per Person, 1961–2002

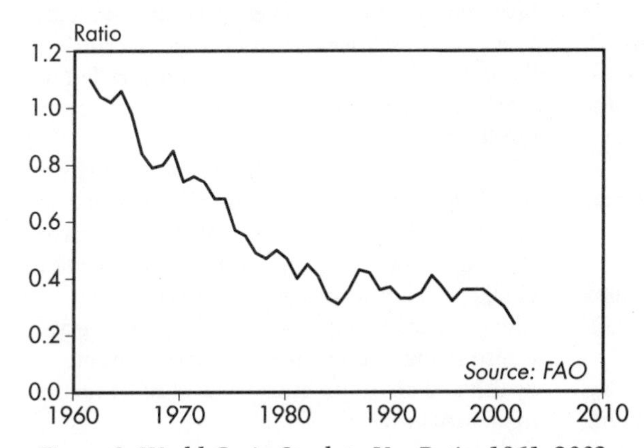

Figure 3: World Grain Stock-to-Use Ratio, 1961–2002

World Grain Production, 1961–2002

Year	Total (million tons)	Per Person (kilograms)
1961	805	261
1962	858	274
1963	867	270
1964	914	279
1965	914	273
1966	992	290
1967	1032	296
1968	1065	299
1969	1073	295
1970	1087	293
1971	1194	315
1972	1156	299
1973	1246	316
1974	1216	303
1975	1241	304
1976	1348	324
1977	1333	315
1978	1454	338
1979	1413	323
1980	1418	318
1981	1496	330
1982	1552	337
1983	1478	315
1984	1632	342
1985	1665	343
1986	1678	340
1987	1618	322
1988	1565	307
1989	1700	328
1990	1779	337
1991	1717	320
1992	1797	330
1993	1726	313
1994	1777	317
1995	1714	302
1996	1882	326
1997	1902	325
1998	1903	321
1999	1890	315
2000	1863	307
2001	1894	308
2002 (prel)	1833	294

Source: U.N. Food and Agriculture Organization.

Meat Production and Consumption Grow

Danielle Nierenberg

The world's appetite for meat continues to grow, with 242 million tons produced in 2002—an increase of 2.5 percent from 2001.[1] (See Figure 1.) Meat production has doubled since 1977, and over the last half-century it has increased fivefold.[2] Production of beef, poultry, pork, and other meats has risen to nearly 40 kilograms per person, more than twice as much as was available in 1950.[3] (See Figure 2.)

Consumers in industrial nations eat more than 80 kilograms of meat per person, most of it from pork and poultry, compared with just 28 kilograms for people in developing countries.[4] In fact, people in industrial nations eat three to four times as much meat as people living in developing countries.[5]

Yet two thirds of the gains in meat consumption in 2002 occurred in developing countries, where urbanization, rising incomes, and the globalization of trade are changing diets and increasing per capita consumption of meat.[6] And as developing countries climb up the "protein ladder," they have overtaken industrial nations as meat producers by accounting for 56 percent of production—an increase of 5 percent since 1995.[7]

LINKS pp. 28, 96

Pork production reached over 93 million tons in 2002, followed by poultry production (72 million tons), and beef (60 million tons).[8] Other types of meat, including sheep and goat meat, accounted for 16 million tons of the total output.[9] (See Figure 3.)

Pigs dominate meat production and consumption in China—half of the world's pigs are raised and eaten there.[10] The United States produces and consumes the most poultry in the world, and Brazil is the world's largest producer of beef and its second-largest consumer, behind only the United States.[11]

Since the early 1960s, the number of livestock has increased 60 percent, from 3 billion to more than 5 billion, and the number of fowl has quadrupled from 4 billion to 16 billion.[12] Industrial feedlots are the most rapidly growing production system for these animals, producing 43 percent of the world's beef and more than half of the world's pork and poultry.[13] These "factory farms" are also responsible for huge amounts of manure and air pollution and for the overuse of antibiotics as crowded conditions encourage the rapid spread of disease.

Producing meat requires large amounts of grain—most of the corn and soybeans harvested in the world are used to fatten livestock.[14] Producing 1 calorie of flesh (beef, pork, or chicken) requires 11–17 calories of feed. So a meat eater's diet requires two to four times more land than a vegetarian's diet.[15] Soybeans, wheat, rice, and corn also produce three to eight times as much protein as meat.[16]

The U.N. Food and Agriculture Organization predicts that meat production will grow to more than 300 million tons by 2020.[17] Environmental and health concerns could be a constraint on that, however. Manure from hog factories, chicken houses, and feedlots for cattle can contaminate groundwater and rivers and can pollute the air.[18] Cattle also contribute to climate change by emitting methane gas, and overgrazing has decimated once fertile and productive grasslands from Africa to Latin America.[19]

Meat recalls, foot-and-mouth disease, and mad cow disease (BSE—bovine spongiform encephalopathy) have increased concerns about the safety of eating meat. During the summer of 2002, millions of pounds of contaminated beef and other meat products were recalled by the U.S. government.[20] In Japan, beef consumption has been declining since the first case of BSE was reported there in 2001.[21] Concerns over drug residues in poultry led to market closures for U.S.-produced chicken in the Russian Federation.[22]

In the United States, high rates of obesity, heart disease, cancer, and other diseases associated with high-fat, high-cholesterol diets have led some people to shun red meat in favor of chicken and others to give up meat entirely. The popularity of grass-fed and organic meats is also rising as consumers realize the high health and environmental costs of meat raised in factory farms.[23]

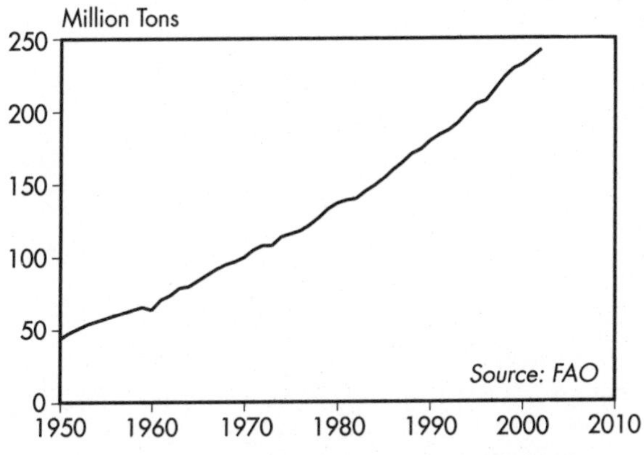

Figure 1: World Meat Production, 1950–2002

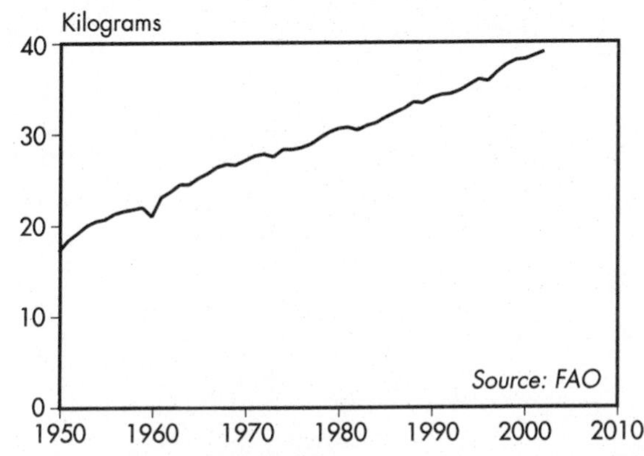

Figure 2: World Meat Production Per Person, 1950–2002

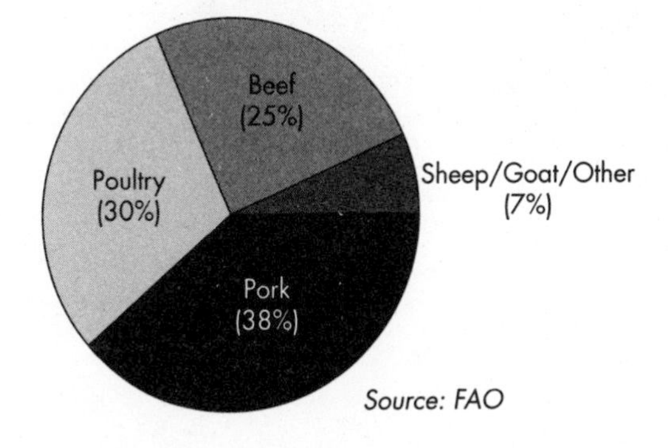

Figure 3: World Meat Production by Source, 2002

World Meat Production, 1950–2002

Year	Total	Per Person
	(million tons)	(kilograms)
1950	44	17.2
1955	58	20.7
1960	64	21.0
1965	84	25.2
1970	100	27.1
1971	105	27.6
1972	108	27.8
1973	108	27.5
1974	114	28.3
1975	116	28.3
1976	118	28.5
1977	122	28.9
1978	127	29.6
1979	133	30.2
1980	137	30.6
1981	139	30.7
1982	140	30.4
1983	145	30.9
1984	149	31.2
1985	154	31.8
1986	160	32.3
1987	165	32.8
1988	171	33.5
1989	174	33.4
1990	180	34.0
1991	184	34.3
1992	187	34.4
1993	192	34.8
1994	199	35.4
1995	205	36.0
1996	207	35.8
1997	215	36.8
1998	223	37.6
1999	229	38.1
2000	232	38.2
2001	237	38.6
2002 (prel)	242	39.0

Source: U.N. Food and Agriculture Organization.

Energy and Atmosphere Trends

© LM Glasfiber, 2001

Fossil Fuel Use Up

Nuclear Power Rises

Wind Power's Rapid Growth Continues

Carbon Emissions and Temperature Climb

Fossil Fuel Use Up

Janet L. Sawin

The global use of fossil fuels (coal, oil, and natural gas) increased by 1.3 percent in 2002, to 8,034 million tons of oil equivalent, according to preliminary estimates based on government and industry sources.[1] (See Figure 1.) This compares with a 0.3-percent rise in 2001.[2] Fossil fuel consumption was 4.7 times the level in 1950, and it now accounts for 77 percent of world energy use.[3]

Global oil use rose by just 0.5 percent in 2002, due in part to a sluggish global economy, according to International Energy Agency (IEA) early estimates.[4] (See Figure 2.) The United States, which uses about 26 percent of global oil, saw only a slight increase in demand.[5] And oil use fell in Europe by an estimated 0.7 percent.[6] It also declined by 0.6 percent (combined average) in Japan, South Korea, Australia, and New Zealand and by 2.6 percent in Latin America.[7] Growth was strongest in China, where demand was up 5.7 percent, followed by the Middle East (2.5 percent) and the former Soviet bloc (1.9 percent).[8]

LINKS p. 40

After a brief but steep decline in the late 1990s, coal use is again on the rise. In 2002, global coal consumption was an estimated 2,298 million tons of oil equivalent—1.9 percent above the 2001 figure.[9] In the United States, which uses nearly 25 percent of the world's coal, demand fell by about 0.5 percent.[10] But China, accounting for 23 percent of global coal use, saw an increase of around 4.9 percent—a sharp rebound following declines in the late 1990s.[11] While China has banned coal burning in some regions with smog and acid rain problems, output from state mines has increased recently.[12]

Natural gas consumption grew by 2 percent, to 2,207 million tons of oil equivalent.[13] The United States, which consumes about 27 percent of the world's natural gas, used 3.7 percent less during the first 10 months of 2002 compared with the same period in 2001.[14] The decline was due primarily to mild winter weather early in the year. Among industrial nations as a whole, natural gas use fell 2.4 percent through November, with the greatest drop in Japan (down 10.4 percent) and the highest increase in Norway (up 81 percent).[15]

Globally, however, natural gas has become the fastest growing of the fossil fuels, and represents an increasing share of global energy use. Today natural gas accounts for nearly 24 percent of world energy consumption, compared with 22.5 percent a decade ago.[16] The increase is due to a number of factors, including an abundance of gas supplies in many countries and the lower environmental impacts of gas use compared with the other fossil fuels.[17] Much of the recent rise in gas use and the projected future increase result from efforts to reduce emissions of air pollutants—primarily through switching from coal and oil to gas in power plants.[18]

In the short term, major uncertainties remain in assessing future trends for fossil fuel use, including the potential economic and political consequences of turmoil in the Middle East. The return of El Niño in late 2002 and early 2003 will likely alter rainfall patterns and bring more extreme temperatures, affecting hydropower production and natural gas demand as weather patterns shift.

For the longer term, the International Energy Agency projects that global primary energy demand will increase 1.7 percent annually between 2000 and 2030, reaching 15,300 million tons of oil equivalent in 2030.[19] Fossil fuels are expected to meet more than 90 percent of the increased demand, with most of this growth occurring in the developing world.[20] But even with this rapid growth, the IEA projects that 18 percent of the people in the world in 2030 will still lack access to modern energy services such as electricity.[21]

Yet the IEA forecast is based on assumptions that are tenuous at best. It assumes that prices for most fuels will remain virtually unchanged through 2010 and that energy taxes will not be modified. It also assumes that global oil production will continue to rise, despite the fact that many analysts project it will peak prior to 2020.[22] While oil consumption is likely to be limited by geological and political constraints, combustion of coal will probably be limited by its associated health and environmental costs, particularly global climate change.

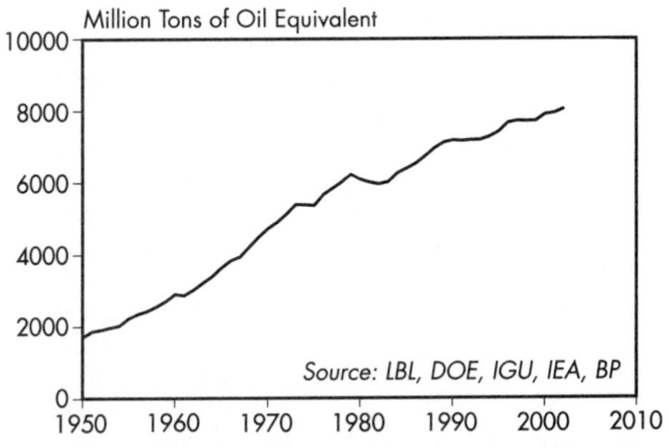

Figure 1: World Fossil Fuel Consumption, 1950–2002

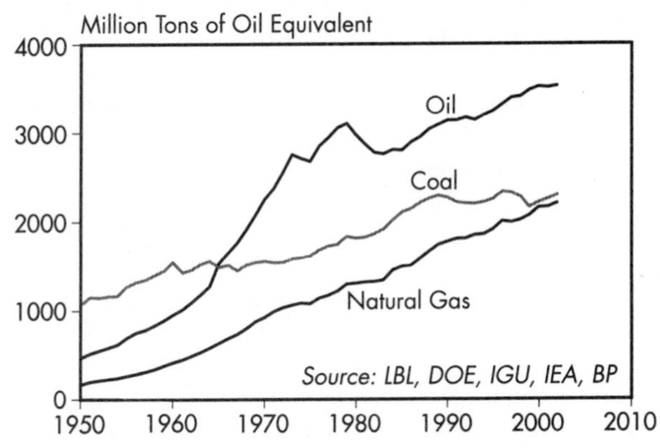

Figure 2: World Fossil Fuel Consumption by Source, 1950–2002

World Fossil Fuel Consumption, 1950–2002

Year	Coal	Oil	Natural Gas
	(million tons of oil equivalent)		
1950	1,074	470	171
1955	1,270	694	266
1960	1,544	951	416
1965	1,486	1,530	632
1970	1,553	2,254	924
1971	1,538	2,377	988
1972	1,540	2,556	1,032
1973	1,579	2,754	1,059
1974	1,592	2,710	1,082
1975	1,613	2,678	1,075
1976	1,681	2,852	1,138
1977	1,726	2,944	1,169
1978	1,744	3,055	1,216
1979	1,834	3,103	1,295
1980	1,814	2,972	1,304
1981	1,826	2,868	1,318
1982	1,863	2,776	1,322
1983	1,914	2,761	1,340
1984	2,011	2,809	1,451
1985	2,107	2,801	1,493
1986	2,143	2,893	1,504
1987	2,211	2,949	1,583
1988	2,261	3,039	1,663
1989	2,293	3,088	1,738
1990	2,270	3,136	1,774
1991	2,218	3,138	1,806
1992	2,204	3,170	1,810
1993	2,200	3,141	1,849
1994	2,219	3,200	1,858
1995	2,255	3,247	1,914
1996	2,336	3,323	2,004
1997	2,324	3,396	1,992
1998	2,280	3,410	2,017
1999	2,163	3,481	2,069
2000	2,217	3,519	2,158
2001	2,255	3,511	2,164
2002 (prel)	2,298	3,529	2,207

Source: Worldwatch estimates based on BP, DOE, IEA, IGU, and LBL.

Nuclear Power Rises

Nicholas Lenssen

Between 2001 and 2002, total installed nuclear power generating capacity increased by more than 5,000 megawatts (nearly 1.5 percent), the fastest percentage growth for the industry since 1993.[1] (See Figure 1.) Seven new reactors were grid-connected in 2002—four in China, two in South Korea, and one in the Czech Republic—bringing the world's total to 437.[2] Additional capacity increases in 2002 came from squeezing more power from existing reactors.

In 2002, construction started on six new reactors, all in India.[3] (See Figure 2.) Some 25 reactors remain under active construction (with a combined capacity of 20,959 megawatts), the fewest since the 1960s.[4] And seven reactors were permanently closed, bringing the total number of retired reactors to 106 (representing 31,439 megawatts).[5] (See Figure 3.)

China expects to complete construction on another four reactors in the next few years, with plans to quadruple existing capacity to 20,000 megawatts by 2020.[6] Likewise, India's recent surge in new construction will increase the country's nuclear generating capacity by nearly 150 percent, to 6,000 megawatts. India's goal for 2020 is also 20,000 megawatts of nuclear capacity.[7]

In some traditional markets, however, nuclear power is facing tougher times. The privatized English nuclear company, British Energy, was saved from receivership last year by a government bailout.[8] Nuclear power simply proved too costly for the competitive U.K. electricity market. The company moved to sell its stakes in U.S. and Canadian nuclear energy companies, perhaps frustrating efforts to restart two mothballed reactors in Canada.[9]

Belgium's government furthered its effort to phase out nuclear power by 2025 with a parliamentary vote of approval.[10] The Finnish parliament, in contrast, narrowly voted in favor of building the country's fifth reactor.[11] Still, not a single new reactor is under construction in Western Europe.

In Eastern Europe, Romania moved closer to obtaining financing to restart construction on its second reactor, which began in 1982.[12] Russia continues to have two reactors under construction, but Ukraine made little progress in obtaining resources to restart mothballed projects.[13]

In the United States, the biggest owner of nuclear power, Exelon Corp., dropped out of an international consortium to develop a new, smaller power plant.[14] Despite press reports that utilities planned to build new reactors, no serious movement in that direction occurred as the U.S. electric industry reeled from financial losses, scandals, and generation overcapacity.[15]

Japan's nuclear industry suffered "the most serious setback ever in public trust in the country's nuclear power program" in 2002: it was disclosed that the country's largest utility and nuclear operator, Tokyo Electric (Tepco), had been systematically falsifying safety inspections since the late 1980s.[16] As government regulators scrambled to close Tepco's reactors, the scandal spread to other utilities, where similar falsification became evident.[17]

The cover-up had immediate consequences for Japan's industry, which only has three reactors under construction. The governor of Fukui withdrew permission to build two new reactors in his province, and two other governors rescinded their consent to load mixed plutonium/uranium fuel in existing reactors as planned.[18]

South Korea completed two more reactors in 2002, leaving just two under construction there at year's end. But the country anticipates completing eight more reactors in the next 11 years.[19]

Concerns over the proliferation of nuclear weapons rose significantly in 2002, colliding with plans to build nuclear power reactors in both Iran and North Korea. The United States accused Iran of working on nuclear weapons, although it failed to convince Russia to stop helping Iran finish at least two reactors.[20] Iran says it is preparing for 6,000 megawatts of new power reactors.[21]

Meanwhile, the fate of two reactors to be built in North Korea was unclear after the country admitted having a secret nuclear weapons program.[22] The reactors, first promised in 1994 by South Korea, Japan, and the United States in exchange for ending a weapons program, were in the early stages of construction as of late 2002.

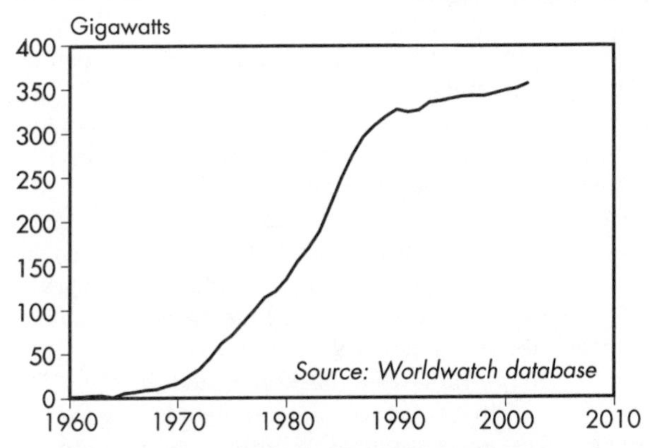

Figure 1: World Electrical Generating Capacity of Nuclear Power Plants, 1960–2002

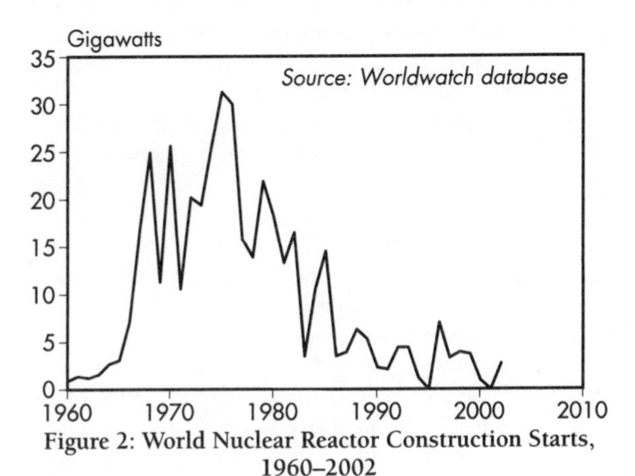

Figure 2: World Nuclear Reactor Construction Starts, 1960–2002

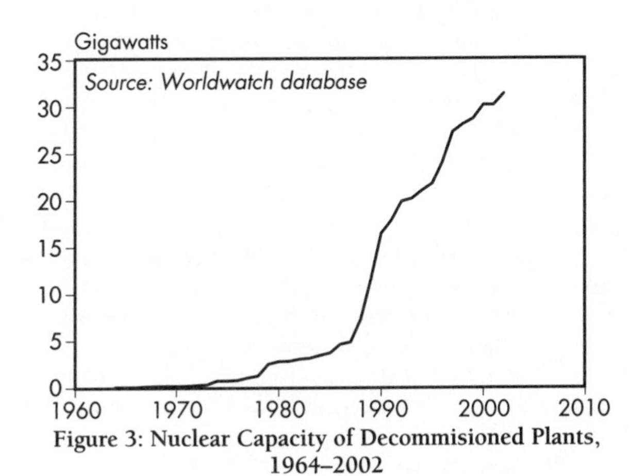

Figure 3: Nuclear Capacity of Decommisioned Plants, 1964–2002

World Net Installed Electrical Generating Capacity of Nuclear Power Plants, 1960–2002

Year	Capacity (gigawatts)
1960	1
1965	5
1970	16
1971	24
1972	32
1973	45
1974	61
1975	71
1976	85
1977	99
1978	114
1979	121
1980	135
1981	155
1982	170
1983	189
1984	219
1985	250
1986	276
1987	297
1988	310
1989	320
1990	328
1991	325
1992	327
1993	336
1994	338
1995	340
1996	343
1997	343
1998	343
1999	346
2000	349
2001	352
2002 (prel)	357

Source: Worldwatch Institute database, compiled from the IAEA and press reports.

Wind Power's Rapid Growth Continues

Janet L. Sawin

Wind energy generating capacity reached nearly 32,000 megawatts by the end of 2002, an increase of 27 percent over 2001.[1] (See Figure 1.) Spurred on by falling costs, concern about climate change, and new government policies, wind remains the fastest-growing energy source in the world. Global wind capacity has tripled since 1998.[2] In early 2002, wind power provided enough electricity to meet the residential electricity needs of 35 million people worldwide.[3] Many more people get at least some of their electricity from the wind.

LINKS p. 40

Global capacity net additions in 2002 totaled approximately 6,720 megawatts—a new record.[4] (See Figure 2.) Yet the rate of growth was slower than expected due to a slump in the U.S. market, which continues to swing widely in response to short-term extensions of a federal wind energy tax credit. The United States installed only 410 megawatts of new capacity in 2002, compared with 1,714 megawatts in 2001, bringing its total to 4,685 megawatts.[5] But up to 1,800 more megawatts are expected in 2003 as developers rush to install projects before the tax credit expires at year's end.[6]

Europe installed an estimated 5,870 megawatts of capacity in 2002, 31 percent more than in 2001.[7] Europe has nearly 73 percent of global wind capacity—thanks to strong, consistent policies driving demand for renewable energy technologies, particularly in Germany, Spain, and Denmark, which accounted for 90 percent of the capacity installed in Europe during 2002.[8]

In fact, more than half of Europe's and 38 percent of the world's wind capacity is found in Germany.[9] In 2002, Germany set another record, adding 3,250 megawatts to end the year with just over 12,000 megawatts of total capacity—enough to provide 4.7 percent of the nation's electricity.[10] In October, Chancellor Schröder announced plans to reduce Germany's greenhouse gas emissions 40 percent by 2020.[11] Wind power will play a large role in this plan.

Spain experienced another strong growth year as well, adding 1,490 megawatts for a total of 4,830 megawatts, surpassing the United States to rank second worldwide.[12] This is an impressive accomplishment, given that Spain's wind industry is not yet a decade old.

Denmark, a nation of just 5 million people, also installed more wind capacity than the United States. With the addition of nearly 500 megawatts, Denmark ended 2002 with about 2,880 megawatts, enough to generate 21 percent of the country's electricity.[13] Much of this new capacity is operating offshore, thanks to completion of the Horns Rev 160-megawatts project, the world's largest offshore wind farm.[14]

The United Kingdom, despite having the best wind resources in Europe, continues to experience slow growth. It ended 2002 with about 556 megawatts of wind capacity, a 31-megawatt increase.[15] The future looks brighter, however, as planning permission has been granted for the next 450 turbines onshore and the first 90 for offshore use.[16]

Italy added 100 megawatts to maintain its position of sixth overall, ending the year with about 800 megawatts.[17] The Netherlands added 217 megawatts, for a total of 740 megawatts.[18] And three new markets emerged—in Norway (added 80 megawatts), Poland (30 megawatts), and Latvia (21 megawatts).[19]

Beyond Europe and the United States, the most significant growth was in Asia. India added 250 megawatts, keeping it in fifth place with 1,702 megawatts, while Japan's capacity rose 36 percent to 486 megawatts.[20] Although China's market appears to have slowed in 2002, rising by 16 percent to about 470 megawatts, that country has more than 1,800 megawatts of wind capacity in the development pipeline.[21] Future growth in Japan and China is expected to be rapid, as even offshore wind is now cost-competitive with many conventional energy options.[22]

The global wind industry currently employs about 100,000 people.[23] Most of the jobs are in Europe, and European companies manufacture 80 percent of all wind turbines sold worldwide.[24] The global large-turbine market is expected to surpass $16 billion annually by 2007.[25] The investment firm Merrill Lynch projects that wind power will grow 15-fold over the next 20 years.[26]

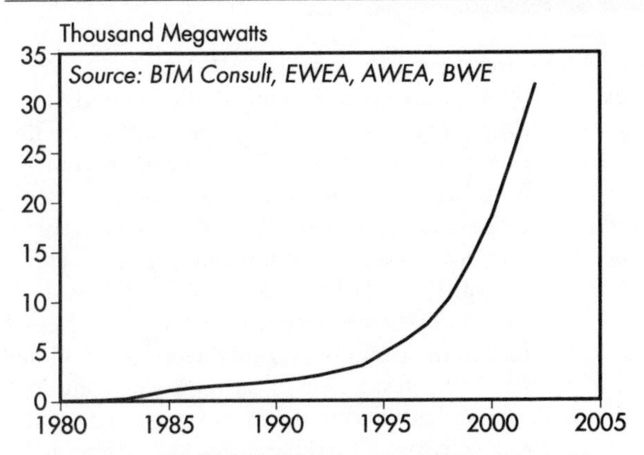

Figure 1: World Wind Energy Generating Capacity, 1980–2002

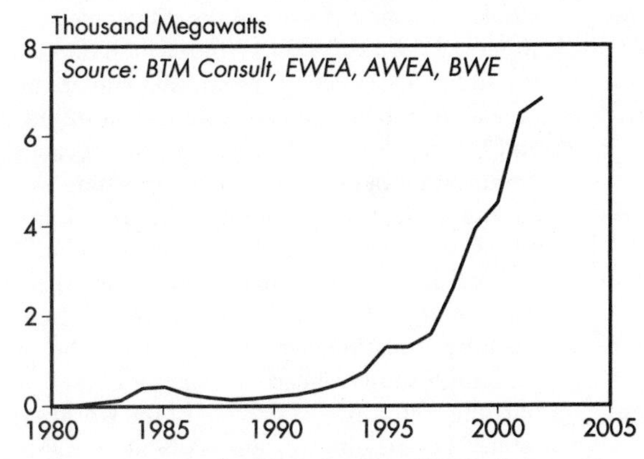

Figure 2: Annual Addition to World Wind Energy Generating Capacity, 1980–2002

World Wind Energy Generating Capacity, Total and Net Annual Additions, 1980–2002

Year	Total	Annual Addition
	(megawatts)	
1980	10	5
1981	25	15
1982	90	65
1983	210	120
1984	600	390
1985	1,020	420
1986	1,270	250
1987	1,450	180
1988	1,580	130
1989	1,730	150
1990	1,930	200
1991	2,170	240
1992	2,510	340
1993	2,990	500
1994	3,490	730
1995	4,780	1,290
1996	6,070	1,290
1997	7,640	1,570
1998	10,150	2,510
1999	13,930	3,780
2000	18,450	4,520
2001	24,930	6,480
2002 (prel)	31,650	6,720

Source: BTM Consult, EWEA, AWEA, and BWE.

Carbon Emissions and Temperature Climb

Molly O. Sheehan

Global average temperature climbed to 14.52 degrees Celsius in 2002, supplanting 2001 as the second hottest year since recordkeeping began in the late 1800s, according to the Goddard Institute for Space Studies.[1] (See Figure 1.) Other centers of climate analysis, using roughly the same network of land and sea temperature gauges, also rank 2002 as second only to 1998 in warmth, and find that the nine warmest years have occurred since 1990.[2]

Scientists have linked the warming trend that took off in the twentieth century to the buildup of carbon dioxide (CO_2) and other heat-trapping gases.[3] By burning fossil fuels, people released some 6.44 billion tons of carbon into the atmosphere in 2002, a 1-percent increase over the previous year, raising atmospheric CO_2 concentration to 372.9 parts per million by volume.[4] (See Figure 2.)

LINKS pp. 34, 56, 82, 84, 92

Measurements taken at the Mauna Loa Observatory in Hawaii show an 18-percent increase in CO_2 levels from 1960 to 2002.[5] Scientists estimate that levels have risen 31 percent since the onset of the Industrial Revolution around 1750.[6] The current concentration has not been exceeded in at least 420,000 years—and likely in 20 million years.[7]

Oscillations in the temperature of the tropical Pacific Ocean are linked to atmospheric CO_2 levels as well as to year-to-year fluctuations in temperature.[8] The world's oceans, which contain about 50 times as much CO_2 as the atmosphere does, are able to take up more carbon when cool.[9] When the sea surface warms in the equatorial Pacific, as it does during an El Niño event, the ocean absorbs less carbon, so atmospheric CO_2 levels rise, along with global temperature.[10]

In May 2002, ocean buoys in the central Pacific started reading warmer-than-average temperatures, heralding the onset of El Niño, which persisted into 2003, sharply changing patterns of rainfall, temperature, and winds in some regions and contributing to, for instance, droughts in India, Australia, and Africa and floods in Europe.[11] Scientists believe that this El Niño may help push global average temperature to a new high in 2003.[12]

Indicators of a warming world abound. Biologists are recording spring events such as the first flowering of plants and the arrival of migrant birds occurring earlier, and are finding the geographic ranges of birds, butterflies, and herbs moving poleward.[13] Mountaintop glaciers are retreating in Alaska, Asia, the Alps, Indonesia, Africa, and South America.[14] Global sea levels rose in the twentieth century about 1–2 millimeters a year, faster than in the nineteenth century.[15]

Poor nations are the most vulnerable to climate change. As temperatures have risen on mountaintops in Rwanda and in other African highlands, malaria-carrying mosquitoes have extended their range, infecting more people.[16] Cholera bacteria thrived in the warm ocean waters of the 1997–98 El Niño, which flooded the Indian Ocean coast, prompting cholera outbreaks in Djibouti, Somalia, Kenya, Tanzania, and Mozambique.[17] Over the last two decades, floods and other weather-related disasters were among the factors prompting some 10 million people to migrate from Bangladesh to India.[18]

Wealthy nations contribute the most to climate change. With less than 5 percent of the world's population, the United States is the single largest source of carbon from fossil fuel burning—emitting 24 percent of the world's total.[19] Per person, U.S. emissions are roughly double that of other major industrial nations and 17 times that of India.[20] (See Figure 3.) China, home to one fifth of the world's people, ranks a distant second to the United States in emissions, with just 12 percent of the total.[21]

Some progress toward reducing global carbon emissions was made in 2002, when Japan, Canada, and the 15 nations of the European Union ratified the 1997 Kyoto Protocol on climate change.[22] For the protocol to come into force, 55 nations representing 55 percent of the 1990 emissions of industrial and former Eastern bloc nations must ratify it. By the end of January 2003, 104 nations representing 44 percent of emissions had done so.[23] As the United States and Australia have pulled out of the process, Russia must ratify the protocol for it to come into force.[24]

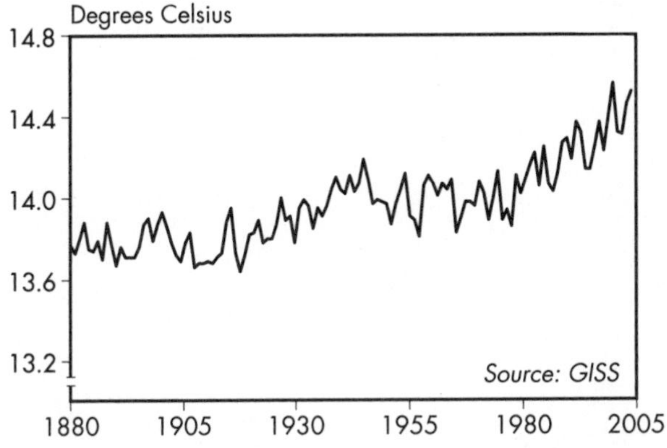

Figure 1: Global Average Temperature at Earth's Surface, 1880–2002

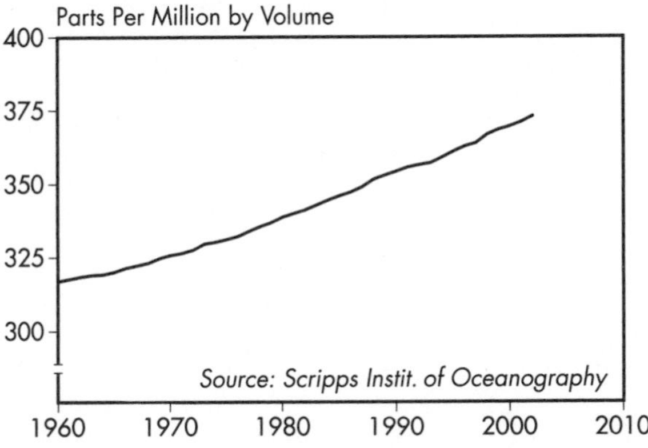

Figure 2: Atmospheric Concentrations of Carbon Dioxide, 1960–2002

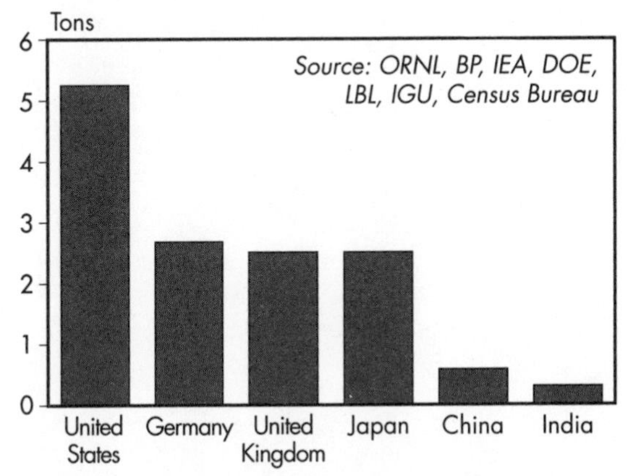

Figure 3: Carbon Emissions Per Person in Selected Countries, 2002

Global Average Temperature and Carbon Emissions from Fossil Fuel Burning, 1950–2002, and Atmospheric Concentrations of Carbon Dioxide, 1960–2002

Year	Temperature	Emissions	Carbon Dioxide
	(degrees Celsius)	(mill. tons of carbon)	(parts per mill. by vol.)
1950	13.87	1612	n.a.
1955	13.89	2013	n.a.
1960	14.01	2535	316.7
1965	13.9	3087	319.9
1970	14.02	3997	325.5
1975	13.94	4518	331.0
1976	13.86	4776	332.0
1977	14.11	4910	333.7
1978	14.02	4961	335.3
1979	14.09	5249	336.7
1980	14.16	5177	338.5
1981	14.22	5004	339.8
1982	14.06	4961	341.0
1983	14.25	4944	342.6
1984	14.07	5116	344.2
1985	14.03	5277	345.7
1986	14.12	5439	347.0
1987	14.27	5561	348.7
1988	14.29	5774	351.3
1989	14.19	5882	352.7
1990	14.37	5953	354.0
1991	14.32	6023	355.5
1992	14.14	5907	356.4
1993	14.14	5904	357.0
1994	14.25	6053	358.9
1995	14.37	6187	360.9
1996	14.23	6326	362.6
1997	14.40	6422	363.8
1998	14.56	6407	366.6
1999	14.32	6239	368.3
2000	14.31	6315	369.4
2001	14.46	6378	370.9
2002 (prel)	14.52	6443	372.9

Source: Goddard Institute for Space Studies, ORNL, BP, IEA, DOE, IGU, LBL, and Scripps Instit. of Oceanography.

Economic Trends

Lauren Goodsmith

Economic Growth Inches Up

Foreign Debt Declines

Advertising Spending Stays Nearly Flat

Tourism Growing But Still Shaky

World Heritage Sites Rising Steadily

Economic Growth Inches Up

Erik Assadourian

Gross world product (GWP)—the aggregated estimate of the global output of goods and services—increased 2.5 percent in 2002, to $48 trillion (in 2001 dollars).[1] (See Figure 1.) Although this means the GWP reached another new high, the increase was below the average of 3.9 percent seen over the years since 1950.[2]

The United States, which accounts for 22 percent of the GWP, increased output by 2.2 percent, driven primarily by robust consumer spending that recovered quickly after the terrorist attacks in September 2001.[3] Latin America's product declined by 0.7 percent, primarily due to the economic crisis in Argentina, which in turn reduced investor confidence in the region.[4] Asia's economy grew by 3.8 percent, spurred by global trade, consumer demand in China and South Korea, and the start of a recovery in the information technology sector.[5] In Africa, gross regional product grew by 2.4 percent—just shy of the global average—but per capita growth there was a mere 0.3 percent as population increased by 18 million.[6]

LINKS pp. 46, 48, 50

With the world's population growing by 74 million in 2002, per capita GWP only increased 1.3 percent, to $7,714.[7] Because governments need to expand infrastructure to keep up with growing numbers, the benefits of economic growth are limited by population growth.[8]

In recent years, a growing number of experts have challenged GWP as an accurate measure of economic growth, let alone of progress.[9] First and foremost, GWP is an absolute measure, counting all expenditures as positive contributions, regardless of their worth to society.[10] It also omits key economic sectors, like subsistence farming and household maintenance.[11] As a counter to this, Redefining Progress, a U.S. nongovernmental research group, created the Genuine Progress Indicator (GPI), which subtracts costs to the economy such as traffic, pollution, and crime while adding unaccounted benefits such as unpaid child care and volunteer work. In the United States, per capita GDP grew 77 percent from 1975 to 2000—compared with GPI growth of just 2 percent.[12] (See Figure 2.)

GWP also ignores the environmental costs of economic activities and does not factor in the value of nature's services on which the global economy depends. These services, such as food production, waste treatment, and climate regulation, have been estimated to be worth anywhere from $18 trillion to $62 trillion—roughly the size of the GWP itself.[13] One recent analysis determined that the wealth of several countries has declined even while gross national product has increased, once depletion of natural capital is factored in.[14]

With growing concern about climate change and shrinking natural resources, many observers are questioning whether traditional economic growth can continue to be thought of as a positive. One measure, the "ecological footprint," looks at per capita use of renewable resources and compares this to the capacity of Earth to generate them. This conservative estimate, which does not include the needs of other species, nonrenewable resource use, or pollution, finds that on average each person uses the resources of 2.3 "global hectares" of productive land.[15] Yet there is only an average of 1.9 hectares of productive area available per person globally.[16]

Thus humanity is withdrawing resources 20 percent faster than Earth can renew them (see Figure 3) and is consequently depleting the world's ecological assets.[17] Indeed, studies show that humans have fully exploited or depleted two thirds of ocean fisheries and have transformed or degraded up to half of Earth's land.[18]

Few countries have remained within their respective ecological capacities—let alone within the global average—and many have far exceeded them. The United States, for instance, used up 9.7 hectares worth of resources per person in 1999—45 percent more than the 5.3 hectares available to each citizen.[19] Even without continued population growth, if the world were to consume as much meat and use as much fossil fuels as Americans do, it would need the resources of five Earths.[20]

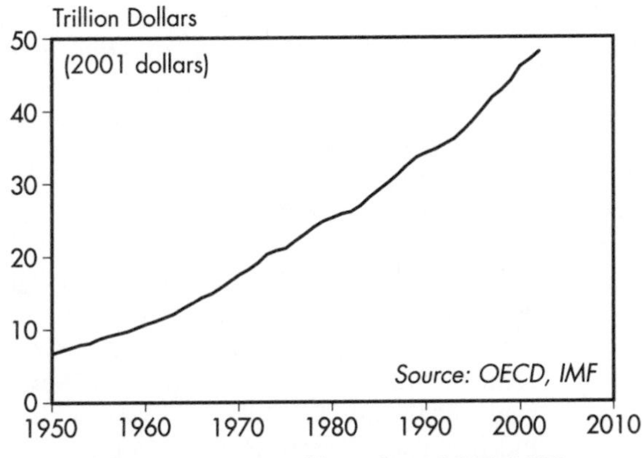

Figure 1: Gross World Product, 1950–2002

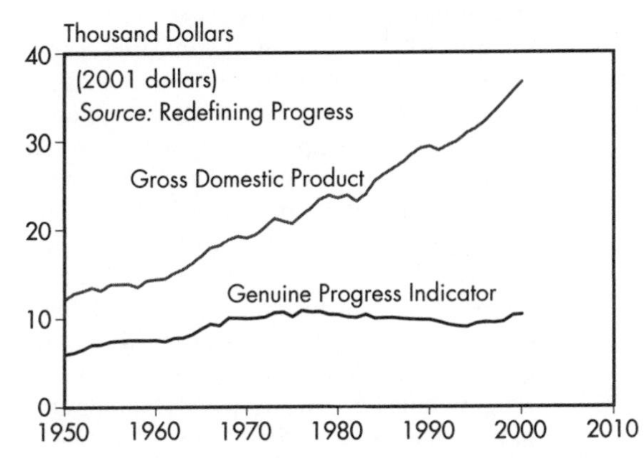

Figure 2: Gross Domestic Product and Genuine Progress Indicator Per Person, United States, 1950–2002

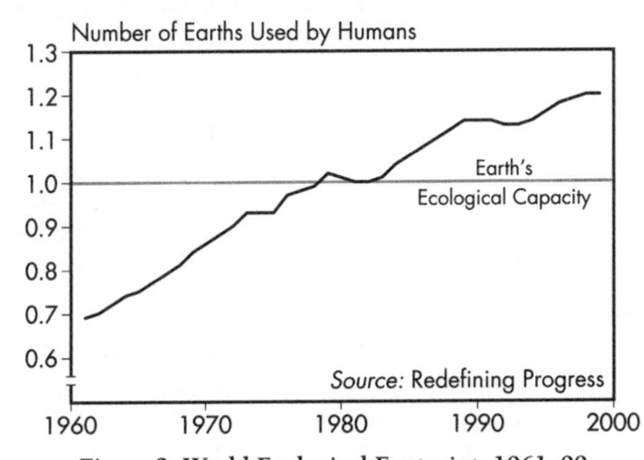

Figure 3: World Ecological Footprint, 1961–99

Gross World Product, 1950–2002

Year	Total (trill. 2001 dollars)	Per Person (2001 dollars)
1950	6.7	2,641
1955	8.7	3,112
1960	10.7	3,516
1965	13.6	4,071
1970	17.5	4,708
1971	18.2	4,805
1972	19.1	4,933
1973	20.3	5,157
1974	20.8	5,174
1975	21.1	5,154
1976	22.1	5,312
1977	23.0	5,432
1978	24.0	5,573
1979	24.8	5,672
1980	25.3	5,688
1981	25.8	5,698
1982	26.1	5,664
1983	26.9	5,728
1984	28.1	5,890
1985	29.1	5,993
1986	30.1	6,101
1987	31.2	6,216
1988	32.5	6,375
1989	33.6	6,470
1990	34.2	6,492
1991	34.7	6,468
1992	35.4	6,499
1993	36.1	6,538
1994	37.3	6,663
1995	38.6	6,791
1996	40.1	6,964
1997	41.7	7,139
1998	42.7	7,202
1999	44.0	7,337
2000	46.0	7,566
2001	46.9	7,617
2002 (prel)	48.0	7,714

Source: Organisation for Economic Co-operation and Development and International Monetary Fund.

Foreign Debt Declines

Molly O. Sheehan

Developing and former Eastern bloc nations borrow money from foreign banks and governments to finance transportation, power generation, schools, loans to local businesses, and other sorts of development projects. In 2001, their cumulative foreign debt shrunk to \$2.44 trillion.[1] (See Figure 1.) More than half of the debt is owed to private, commercial lenders; the rest is owed to national governments, the World Bank, the International Monetary Fund (IMF), and regional development banks.[2]

LINKS pp. 44, 96

Some 78 percent of the debt in 2001 was owed by middle-income nations, which typically pay market-based interest rates and borrow more heavily from commercial than official lenders.[3] The global economic slowdown in 2001 made private banks more averse to risk and less inclined to lend to developing countries, as ratings agencies class some two thirds of developing countries as "speculative-grade" borrowers.[4] New loans to developing nations from private banks fell to \$93 billion in 2001, a 25-percent drop from 2000.[5]

The IMF has proposed a Sovereign Debt Restructuring Mechanism (SDRM) as a bankruptcy process to streamline the restructuring of developing-country debt that would be similar to what is in place within many countries for companies and municipalities.[6] This is intended for commercial debt, so it would be of most use in managing crises in middle-income nations, such as Argentina's economic meltdown in 2001.[7]

While banks and creditor countries have stalled the SDRM, other initiatives are more relevant for low-income nations, which owed roughly 22 percent of outstanding debt in 2001.[8] These nations rely heavily on special loans, some virtually interest-free, from the World Bank and other government agencies. Compared with other regions, sub-Saharan Africa and South Asia owe a greater share of their debt to official lenders.[9] (See Figure 2.)

The total external debt of some nations is higher than they will be able to repay. This "debt overhang" deters foreign investment and drags down the economy, as governments fail to meet people's basic health and education needs.[10] Zambia devoted more than 30 percent of its budget to debt repayments each year in the 1990s, for example, while spending roughly 10 percent on basic social services.[11]

Starting in the late 1980s, through the Paris Club, creditor nations announced a series of special terms for poor nations struggling with high debt—offering longer repayment periods and canceling some debts.[12] Then in 1996 the Group of Seven industrial nations called on the World Bank and the IMF to administer a Heavily Indebted Poor Countries (HIPC) program, which was expanded in 1999, largely in response to pressure from a coalition of nongovernmental organizations called Jubilee 2000.[13]

Some 42 countries, mostly in Africa, can qualify for debt relief after they show a track record of reforms to promote macroeconomic stability and draw up a poverty reduction strategy in consultation with civil society groups.[14] As of January 2003, six nations had completed the program, while another 21 had begun it.[15]

The HIPC relief is worth less than it appears on paper because much of the debt could never be repaid anyway.[16] Many analysts have called for greater debt forgiveness.[17] As corrupt governments have wasted or misappropriated money in the past, and as some of the poorest nations are now overwhelmed by the AIDS crisis, a key challenge for donors and borrowers will be ensuring that funds freed up by debt relief go into areas of urgent need, such as public health and primary education.[18]

Changes in the rules of world trade are needed to help nations service their debts. Poor nations get foreign currency to repay loans through trade, relying heavily on exports of agriculture and textiles, which remain protected in rich nations.[19] The average person in a developing country selling into world markets confronts barriers that are roughly twice as high as those faced by counterparts in industrial nations.[20] In 2002, the U.N. Development Programme called on the World Trade Organization to open up its meetings to counter back-room deals made by a handful of wealthy nations that effectively limit developing nations' power to set trade rules.[21]

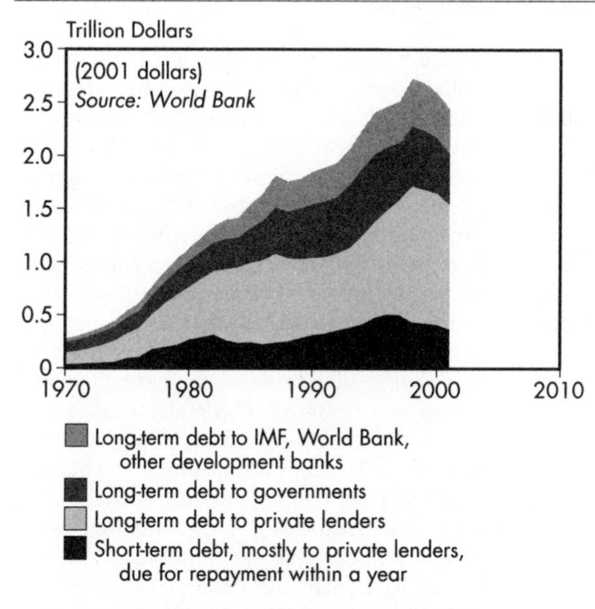

Figure 1: Foreign Debt of Developing and Former Eastern Bloc Nations, 1970–2001

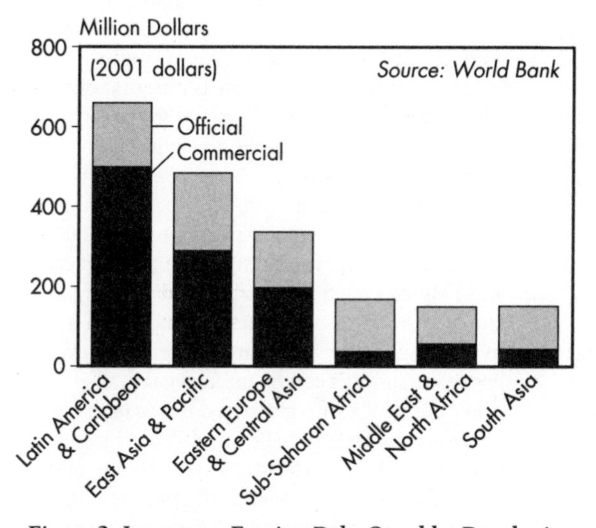

Figure 2: Long-term Foreign Debt Owed by Developing and Former Eastern Bloc Nations to Commercial and Official Lenders, by Region, 2001

Foreign Debt of Developing and Former Eastern Bloc Nations, 1970–2001

Year	Foreign Debt (trillion 2001 dollars)
1970	0.27
1971	0.30
1972	0.34
1973	0.38
1974	0.44
1975	0.54
1976	0.62
1977	0.77
1978	0.90
1979	1.03
1980	1.13
1981	1.23
1982	1.34
1983	1.40
1984	1.41
1985	1.54
1986	1.65
1987	1.81
1988	1.76
1989	1.78
1990	1.85
1991	1.89
1992	1.94
1993	2.07
1994	2.25
1995	2.40
1996	2.46
1997	2.51
1998	2.72
1999	2.68
2000	2.59
2001	2.44

Source: World Bank.

Advertising Spending Stays Nearly Flat

Erik Assadourian

Global advertising expenditures grew 0.6 percent in 2002 to $444 billion; of this total, $309 billion was spent on major media, including television, radio, and newspaper.[1] (See Figure 1.) This modest growth was almost fully driven by the United States, which at $235 billion accounts for over half of the total advertising market.[2] In 2002, U.S. advertising grew by 1.7 percent, stimulated by an economic recovery and cyclical events like the Winter Olympics and the U.S. congressional elections—the latter generating $1 billion in ads.[3] Yet the worldwide increase followed a fall of 9.2 percent in 2001, which was triggered by the U.S. recession, the financial market collapse, the Internet "bubble burst," and terrorist attacks.[4]

LINKS pp. 44, 56, 70

In Japan, which is the second largest advertising market and buys 12 percent of major media advertising, spending fell 5 percent in 2002.[5] In Germany, the third biggest market and the largest one in Europe, spending fell by 6 percent.[6] In contrast, advertising in China, the seventh largest market, is growing quickly; it was unaffected by the downturn in 2001 and has jumped 14 percent over the past two years.[7]

The global average advertising spending per person for 2002 dropped slightly to $71 ($49 spent on major media), as increases in spending were matched by population growth.[8] (See Figure 2.) Yet this figure masks a huge variation across countries. While major media ad spending stood at $4 per person in China and $282 per person in Japan, in the United States it was $494 per person—10 times the global average.[9] (See Figure 3.)

Advertising promotes consumer spending, which in its current form is harming environmental and human well-being. In 2001, for instance, 5 of the top 10 advertisers were car companies.[10] And even while the economy stagnated that year, the global passenger car fleet grew to 523 million, with production of new cars reaching 40 million.[11] Cars burn vast quantities of oil—polluting the air, contributing to respiratory diseases, and stoking climate change.[12]

The pharmaceutical industry, the sixth largest global advertiser, spent $2.5 billion on television and print advertising in 2000 in the United States, directly targeting consumers and generating demand for drugs.[13] While pharmaceuticals can help save lives, advertising can promote unnecessary use of expensive drugs.[14] A recent survey of U.S. physicians found that 92 percent of patients requested an advertised drug from their doctors and that 47 percent of those doctors felt pressured to prescribe those drugs.[15]

Advertising has become pervasive in daily life and continues to expand into new realms. Increasingly, advertisers are marketing to children to shape consumption preferences early and to take advantage of the growing amount of money that people are spending on children, which hit $405 billion in 2000.[16] American children are bombarded with 40,000 television ads per year, up from 20,000 in the 1970s.[17] Half of these encourage children to request unhealthful food and drinks.[18]

In addition, embedded ads, such as product placements in movies, can seriously influence children. In a recent study, researchers found that smoking in movies is strongly associated with youth smoking habits—as strongly as other social influences, such as parental or sibling smoking habits.[19] U.S. advertising to children has spread to schools, where ads adorn walls, sporting equipment, and even educational programming.[20]

To reduce children's exposure to marketing, several countries, including Denmark, Greece, and Belgium, restrict television advertising to children; Sweden and Norway totally ban it.[21] Even full bans are only partly effective, however, because satellites can beam television ads from other countries into restricted markets.[22]

Public interest groups are also working to reduce children's exposure to advertising and to teach children about marketing motives. In the United States, a campaign of the American Legacy Foundation known as The Truth uses controversial ads, education, and grassroots activism to challenge teens not to get manipulated by the tobacco industry's marketing into starting a lethal habit.[23]

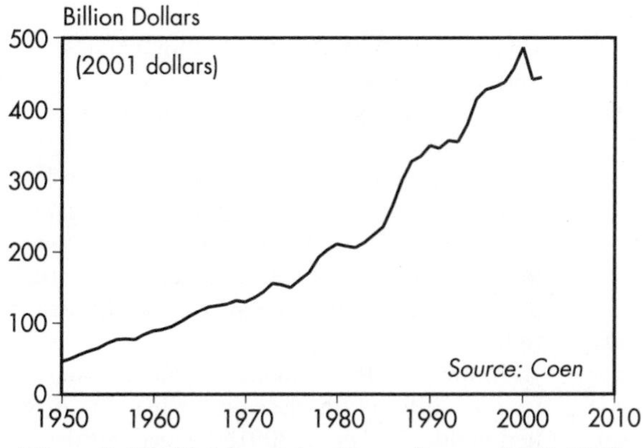

Figure 1: World Advertising Expenditures, 1950–2002

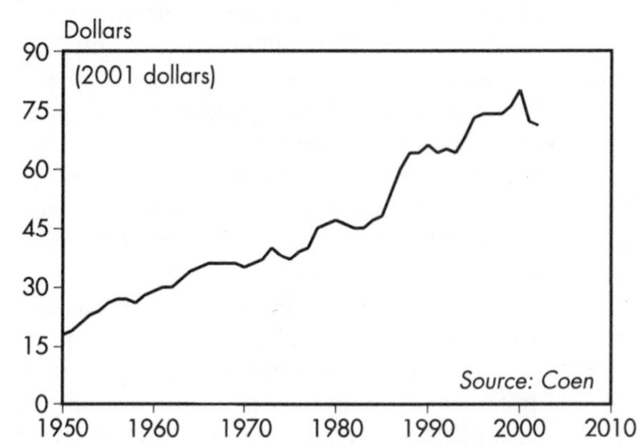

Figure 2: Advertising Expenditures Per Person, 1950–2002

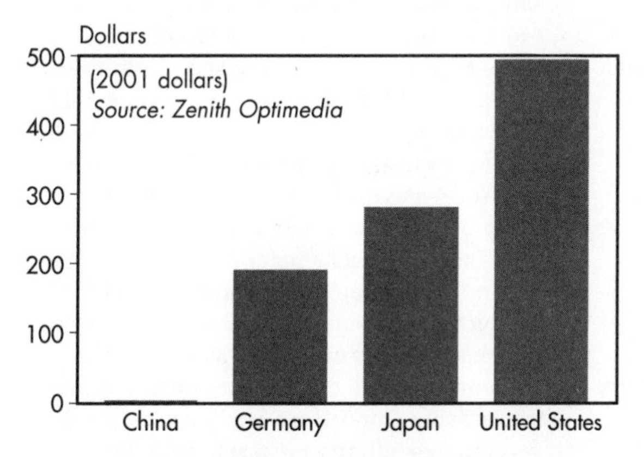

Figure 3: Major Media Advertising Per Person, Selected Countries, 2002

World Advertising Expenditures, 1950–2002

Year	Advertising Expenditures	Expenditures Per Person
	(bill. 2001 dollars)	(2001 dollars)
1950	46	18
1955	72	26
1960	89	29
1965	117	35
1970	130	35
1971	136	36
1972	144	37
1973	156	40
1974	154	38
1975	150	37
1976	161	39
1977	171	40
1978	192	45
1979	203	46
1980	211	47
1981	208	46
1982	206	45
1983	213	45
1984	224	47
1985	235	48
1986	264	54
1987	300	60
1988	327	64
1989	334	64
1990	349	66
1991	345	64
1992	356	65
1993	354	64
1994	378	68
1995	414	73
1996	427	74
1997	431	74
1998	437	74
1999	456	76
2000	486	80
2001	441	72
2002 (prel)	444	71

Source: Coen.

Tourism Growing But Still Shaky

Lisa Mastny

International tourism increased 3 percent in 2002, to 715 million arrivals, according to preliminary estimates by the World Tourism Organization.[1] (See Figure 1.)

This better-than-expected growth came after one of the most difficult episodes in recent tourism history. In 2001, for the first time in nearly 20 years, international tourist arrivals actually declined, by 0.6 percent.[2] The drop reflected the impacts of both the September 11 terrorist attacks in the United States and the global economic slowdown.[3] Receipts from international tourism fell nearly 5 percent in 2001, to $462 billion (in 2001 dollars).[4] (See Figure 2.) (Receipt estimates for 2002 are not yet available.)

LINKS pp. 44, 52

The industry is still reeling from these extensive losses. The World Travel & Tourism Council (WTTC) reports that global tourism employment was down nearly 2 percent in 2002, generating 3 million fewer jobs than two years earlier.[5] And the world's airlines experienced near-zero growth in traffic in 2002, following what the International Air Transport Association has called the "worst year in the history of air transport."[6]

In 2001, the combination of lower passenger travel, a weak global economy, and, in some cases, overambitious expansion strategies contributed to net financial losses of $12 billion for the world's airlines.[7] Many carriers were forced to cut routes, lay off personnel, and restructure operations; in 2002, industry heavyweights US Airways and United Airlines both filed for bankruptcy.[8] Analysts expect brighter news for the airline industry in 2003, the one-hundredth anniversary of aviation.[9]

If global conditions improve, the tourism industry may be well on its way to resuming its strong historic growth.[10] Since 1950, international tourist arrivals have increased nearly 28-fold, growing at an average annual rate of 7 percent.[11]

Europe remains the top destination—capturing 58 percent of arrivals in 2002—though its share of the world's tourists continues to fall from a high of 75 percent in 1964.[12] (See Figure 3.) France was the most visited country in 2002, followed by Spain, the United States, Italy, and China.[13]

For the first year ever, the share of the world's tourists visiting East Asia and the Pacific surpassed the portion visiting the Americas—reaching nearly 18 percent (up from less than 1 percent in 1950).[14] Arrivals to Asia are expected to double within the next decade, and by 2020 the region could attract a quarter of all tourism traffic.[15]

In the past decade alone, China has risen from twelfth to fifth place on the list of most visited nations.[16] And in 2001, it edged out the United Kingdom as the fifth highest earner of tourism receipts worldwide.[17] By 2020, China is predicted to be the top international destination, attracting some 130 million visitors a year.[18]

Overall, tourism-related spending accounted for some $4.2 trillion of global economic activity in 2002 and represented 12 percent of total world exports, according to WTTC.[19] And despite the employment slowdown, the activity generated an estimated 199 million jobs—one in every 13 jobs worldwide.[20]

The growing global dependence on tourism has its downside, however. On average, up to half of all tourism income in developing countries "leaks" out of the destination, with much of it going to industrial nations through foreign ownership of hotels and tour companies.[21] And tourism poses a growing threat to the world's natural areas—from small islands to high peaks and the poles.[22] Tourist transportation and infrastructure, as well as the sheer volume of visitors, can bring serious pollution and habitat destruction.[23]

One promising new trend is "sustainable tourism"—environmentally and socially conscious travel that can help protect natural assets as well as generate local income.[24] The rise in "green" hotels and voluntary codes of conduct for tour operators may lessen the environmental effects of the tourism boom.[25] And ongoing efforts to create a Sustainable Tourism Stewardship Council may help the industry develop new international standards for tourism certification.[26]

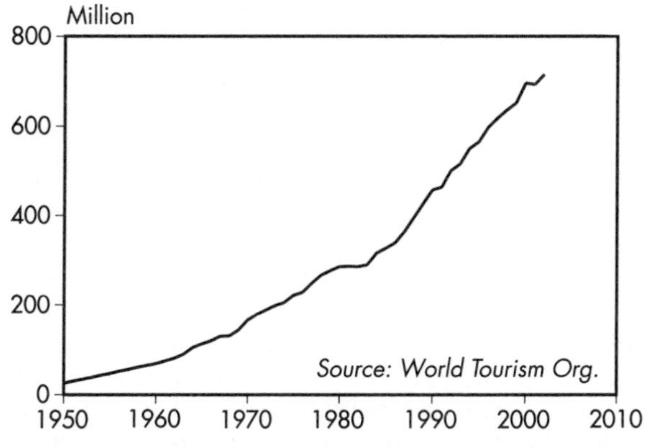

Figure 1: International Tourist Arrivals, 1950–2002

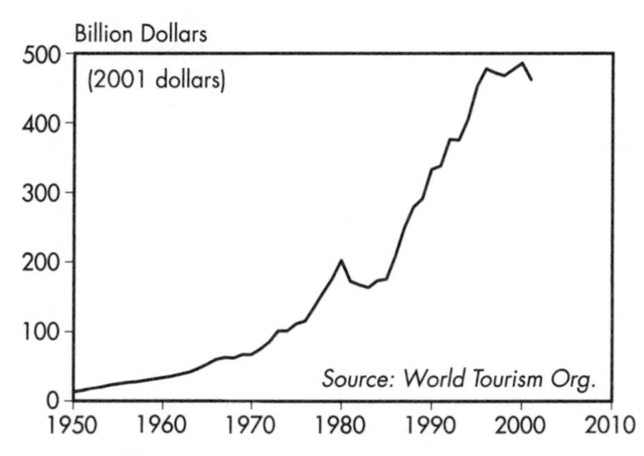

Figure 2: International Tourism Receipts, 1950–2001

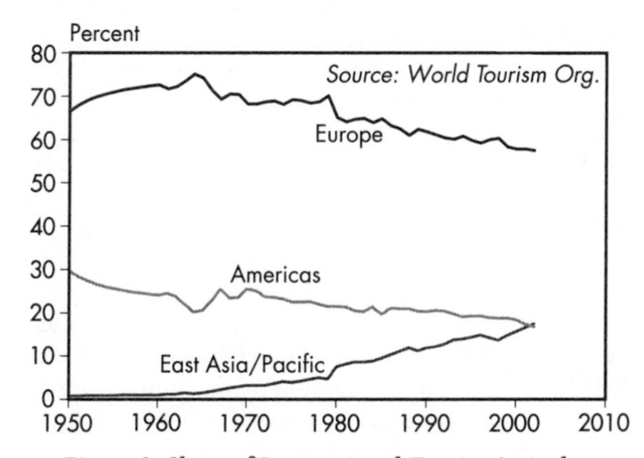

Figure 3: Share of International Tourist Arrivals by Region, 1950–2002

International Tourist Arrivals and Global Tourism Receipts, 1950–2002

Year	Arrivals (million)	Receipts (bill. 2001 dollars)
1950	25	13
1955	47	25
1960	69	34
1965	113	53
1970	166	67
1971	179	75
1972	189	85
1973	199	101
1974	206	101
1975	222	111
1976	229	115
1977	249	135
1978	267	156
1979	277	176
1980	286	202
1981	287	172
1982	286	167
1983	290	163
1984	316	173
1985	327	175
1986	339	208
1987	364	249
1988	395	279
1989	426	291
1990	457	333
1991	463	338
1992	501	376
1993	516	375
1994	550	405
1995	565	453
1996	597	478
1997	618	472
1998	636	468
1999	652	477
2000	696	486
2001	693	462
2002 (prel)	715	–

Source: World Tourism Organization.

World Heritage Sites Rising Steadily

Lisa Mastny

Between 1978 and 2002, the number of World Heritage Sites worldwide increased from 12 to 735.[1] (See Figure 1.) The global total now stands at 730, as several adjacent properties have been merged.[2]

The United Nations Educational, Scientific, and Cultural Organization (UNESCO) confers World Heritage status on cultural or natural sites considered to be of "outstanding value to humanity." [3] The current list includes 563 cultural sites (buildings, monuments, and properties with aesthetic, anthropological, archaeological, ethnological, historical, or scientific value); 144 natural sites (areas with scientific, conservation, or aesthetic value; outstanding physical, biological, and geological formations; or habitats of threatened plant or animal species); and 23 "mixed" sites.[4]

LINKS pp. 50, 74

The properties are located in 125 countries on six continents.[5] Europe, with its heavy concentration of monuments and religious architecture, is home to nearly half of the properties recognized by UNESCO (323 sites), while Asia has nearly a quarter (163 sites).[6] (See Figure 2.) Spain and Italy contain the most individual properties, with at least 35 each, followed by China, France, and Germany.[7] (See Figure 3.)

The idea of protecting the world's shared heritage emerged after World War I in response to growing concern about threats to important cultural and natural landmarks.[8] UNESCO launched the first truly global campaign to save cultural heritage in 1960, when 50 countries raised $80 million to dismantle and rescue the ancient Egyptian temples of Abu Simbel from flooding due to construction of the Aswan High Dam.[9] Other early campaigns focused on conserving Venice in Italy, Pakistan's Bronze Age city of Moenjodaro, and the Buddhist temples of Borobodur in Indonesia.[10]

The formal World Heritage List was established in 1972 following adoption of the Convention Concerning the Protection of the World Cultural and Natural Heritage (the World Heritage Convention).[11] The agreement's 175 signatory countries have pledged to collectively protect natural and cultural areas of "outstanding universal value" that transcend national boundaries and belong to all of humanity.[12]

Earning World Heritage status can be an important way to attract tourist dollars and other resources to national parks, historic landmarks, and other properties. Member countries must individually nominate sites for inclusion on the list. Once a site is approved, governments are encouraged to report on the progress of conservation and to raise public awareness about the property. But not all nominees make the cut: many countries face instability or lack financial or other support, which prevents them from meeting the strict listing requirements.[13] Nevertheless, UNESCO aims to increase the number and diversity of sites, particularly in Africa.[14]

Some existing sites already risk deterioration or disappearance. UNESCO's List of World Heritage in Danger now includes 33 properties, most of which are in Africa.[15] They face a wide range of threats—from armed conflict to abandonment, rampant urban or tourist development, and changes in land use or ownership.[16]

Many properties also face environmental dangers, including pollution, poaching, flooding, and natural disasters like earthquakes, landslides, and volcanic eruptions.[17] Mali's great mosques of Timbuktu are increasingly at risk of encroachment by desert sands, while mining operations threaten both Yellowstone National Park in the United States and the Mount Nimba Strict Nature Reserve spanning Guinea and Côte d'Ivoire.[18]

To date, UNESCO-sponsored campaigns have raised more than $1.5 billion for site preservation worldwide.[19] The organization provides technical assistance and professional training to treaty member countries, as well as emergency assistance to endangered sites.[20]

The world's natural heritage got an additional boost in November 2002 when Conservation International in the United States and the United Nations Foundation announced a three-year, $15-million partnership to support conservation efforts in new and existing properties. Sixteen of the world's 25 so-called biodiversity hotspots are also designated as World Heritage Sites.[21]

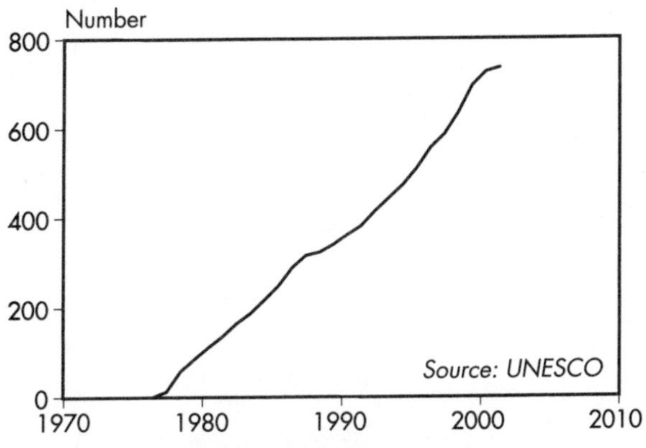

Figure 1: World Heritage Sites, 1978–2002

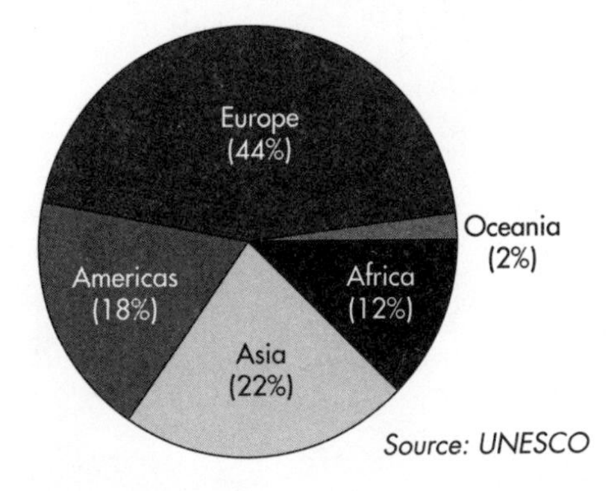

Figure 2: Share of World Heritage Sites by Region, 2002

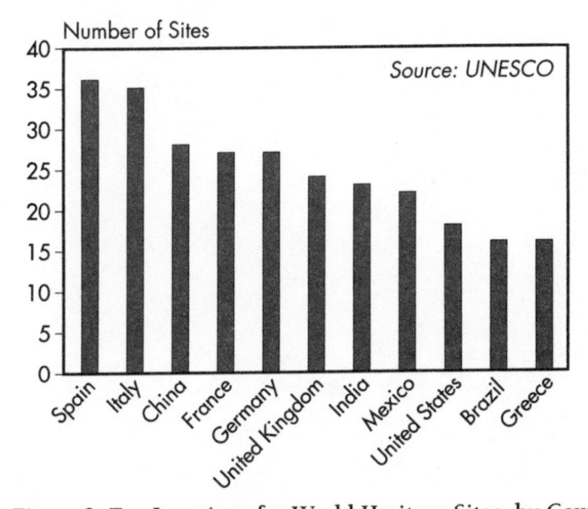

Figure 3: Top Locations for World Heritage Sites, by Country

World Heritage Sites, 1978–2002

Year	Sites (cumulative number)
1978	12
1979	57
1980	85
1981	111
1982	135
1983	164
1984	187
1985	217
1986	248
1987	289
1988	316
1989	323
1990	340
1991	362
1992	382
1993	415
1994	444
1995	473
1996	510
1997	556
1998	586
1999	634
2000	695
2001	726
2002	735

Source: UNESCO.

Transportation and Communications Trends

© Digital Vision

Vehicle Production Inches Up
Bicycle Production Seesaws
Communications Networks Expand
Semiconductor Sales Rebound Slightly

Vehicle Production Inches Up

Michael Renner

Global passenger car production grew 2 percent in 2002, to 40.6 million units.[1] This is still slightly below the 2000 peak output of 41.1 million.[2] Since 1950, annual car production has grown fivefold.[3] (See Figure 1.) Production of sport utility vehicles and other "light trucks" reached a record 15.8 million in 2002, 6 percent more than in 2001.[4] Revised estimates show the global passenger car fleet reaching 531 million in 2002.[5] (See Figure 2.) The United States has one quarter of the cars in the world.[6]

Reflecting continued overcapacity in the industry, passenger car production outpaced sales by almost 2 million vehicles, or more than 5 percent.[7] But light trucks continued to be popular, outrunning production in 2002 by more than 1 million (almost 9 percent).[8] Once primarily used for hauling loads, light trucks are now heavily marketed as passenger vehicles. But even more so than cars, they are increasingly important contributors to air pollution and climate change. In the United States, model-year 2001 light trucks emitted 2.4 times more smog-forming pollutants and 1.4 times more carbon than passenger cars.[9]

LINKS pp. 34, 40, 48

Driving a gasoline-powered car accounts for about 68 percent of the greenhouse gases emitted over the life of the vehicle, but producing and distributing the fuel on which it runs accounts for another 21 percent, while manufacturing the car itself contributes the rest.[10]

Automobile carbon emissions could be reduced significantly by boosting fuel efficiency. Yet fuel economy has remained flat since 1990 in the United States, after substantial improvements since the early 1970s.[11] Efforts to raise mandated fuel efficiency standards failed in the mid-1990s and again in 2002.[12] Carmakers exploit exemptions and loopholes in existing standards, and the Bush administration is considering tax measures that would provide more incentive for buyers to choose the biggest gas-guzzlers.[13]

In 1970, Americans drove some 80 million cars close to 1 trillion miles (almost 1.6 trillion kilometers), burning 5.25 million barrels of fuel per day (mb/d) and emitting 193 million tons of carbon.[14] By 2000, there were about 128 million cars—60 percent more. They traveled 2.3 trillion miles (a growth of 146 percent), consumed 8.2 mb/d of fuel (up 56 percent), and emitted 302 million tons of carbon (also 56 percent more).[15] (See Figure 3.)

In the rest of the world, car density relative to population is much lower than in the United States. In Western Europe and Japan, it is currently comparable to the level the United States reached in the early 1970s; in Eastern Europe, it is similar to that in the 1930s; and in other regions it is even lower.[16]

People outside the United States also use their cars less than Americans. For instance, the average car in the United States travels 10 percent more per year than a car in the United Kingdom, about 50 percent more than one in Germany, and almost 200 percent more than a car in Japan.[17] And Americans drive less fuel-efficient cars, so these figures understate the national differences in gasoline used for driving.[18]

The United States consumed 43 percent of the 19.1 mb/d of world gasoline use for all transportation purposes in 1999.[19] (This number overstates the U.S. share of road fuel use, however, because it does not include diesel fuels, which are popular in Europe.)[20] All in all, the carbon emissions of U.S. automobiles are roughly equivalent to those of the entire Japanese economy—the world's fourth-largest carbon emitter.[21]

The leaders in fuel economy are Honda, Hyundai, Volkswagen, and Subaru. The U.S. "Big Three," by contrast—General Motors, Ford, and Daimler-Chrysler—are among the laggards.[22] One analysis of the six largest carmakers also finds that the Japanese firms Honda, Toyota, and Nissan have a "cleaner" record with regard to smog-forming pollutants than the Big Three.[23]

To date, only Honda and Toyota have introduced "hybrid electric" cars (in which electric power supplements the internal combustion engine, which lowers fuel intake and pollutants).[24] Vehicles running on all different types of alternative power currently account for only a tiny share of the total car fleet. In the United States, fewer than 380,000 such vehicles were on the roads in 2001.[25]

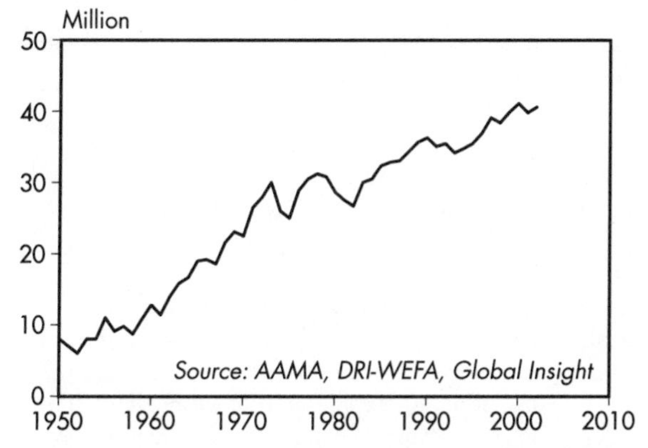

Figure 1: World Automobile Production, 1950–2002

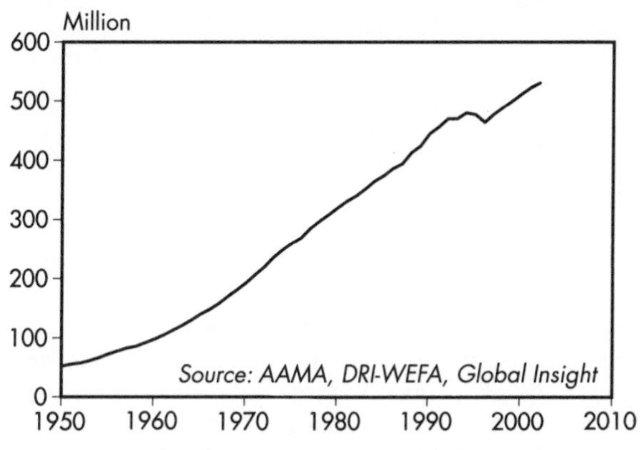

Figure 2: World Passenger Car Fleet, 1950–2002

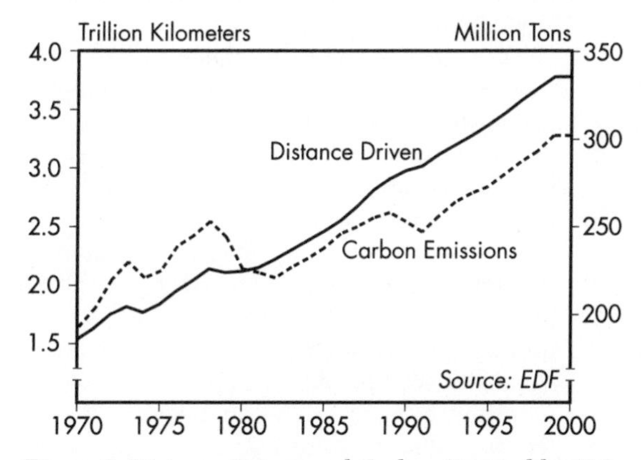

Figure 3: Distance Driven and Carbon Emitted by U.S. Automobiles, 1970–2000

World Automobile Production, 1950–2002

Year	Production (million)
1950	8 0
1955	11.0
1960	12.8
1965	19.0
1970	22.5
1971	26.5
1972	27.9
1973	30.0
1974	26.0
1975	25.0
1976	28.9
1977	30.5
1978	31.2
1979	30.8
1980	28.6
1981	27.5
1982	26.7
1983	30.0
1984	30.5
1985	32.4
1986	32.9
1987	33.1
1988	34.4
1989	35.7
1990	36.3
1991	35.1
1992	35.5
1993	34.2
1994	34.8
1995	35.5
1996	36.9
1997	39.1
1998	38.4
1999	39.9
2000	41.1
2001	39.8
2002 (prel)	40.6

Source: DRI-WEFA, American Automobile Manufacturers Association, and Global Insight.

Bicycle Production Seesaws

Gary Gardner

Global production of bicycles fell by some 7 percent in 2001, to 97 million units, as the production oscillations of the 1990s continued into the new century.[1] (See Figure 1.) And although production data are not yet available for 2002, preliminary indications suggest that the industry's sluggish performance continued.[2]

The fluctuating market is more a reflection of difficulties in inventory management due to globalization of the bicycle industry than it is of changes in demand. As production concentrates in Asia, sellers in distant markets, especially in Europe and the United States, must order stock based on estimates of the strength of their own markets six months or more in advance.[3] Retailers often miscalculate, leaving themselves with burgeoning inventories that are later unloaded, depressing new orders.[4] The result is a seesaw global production cycle in the face of flat global demand.

LINKS p. 56

In the United States, for example, domestic production in 2001 continued its decade-long decline, and imports fell by 19 percent over 2000, yet robust sales were supported by drawing down the millions of bicycles in stock around the country.[5]

The decade-long trend in concentration of production is evident in several ways. Bicycle factories are increasingly rare in the United States, Mexico, and the European Union, as manufacturers move to countries with lower production costs, including Viet Nam and several in Eastern Europe.[6]

China is another major site for new bicycle factories, which has helped consolidate that country's grip on global production. In 2001, China produced 53 percent of the world's bicycles, perhaps the first time ever that one nation has supplied more than half of global output.[7] (See Figure 2.) Increasingly, these bikes are headed overseas: Chinese exports more than doubled between 1997 and 2001, from 14 million units to nearly 35 million.[8] But China also remains the world's leading user of bicycles, despite a steady decline in bicycle use over the past decade.[9]

Chinese inroads into the global market are especially impressive given the barriers to their bicycles that exist in many countries. The European Union and Canada both have stiff import duties on Chinese bikes.[10] The United States, in contrast, does not levy such "dumping" duties.[11] So some 40 percent of Chinese bike exports were shipped to the United States in 2001, accounting for 87 percent of bicycles brought into the country.[12]

The sluggish bicycle market stands in contrast to scattered local interest in promoting more diverse urban transportation systems, including an expanded role for bicycles. The disadvantages of car-centered transportation, including air pollution, sprawl, and congestion, have prompted many cities to rethink their transportation priorities. Programs that restrict the use of private cars for a day are on the rise; some 2,000–3,000 "car-free days" of varying levels of comprehensiveness have been held in the past 10 years.[13] The residents of Bogotá, Colombia, voted overwhelmingly to make that city's February 2000 car-free-day experiment an annual event, and in 2002 eight other Colombian cities restricted car use for a day.[14]

Such initiatives help citizens imagine a transportation system with options other than cars. Many cities build on this conceptual shift by providing bikeways, the physical space needed to make cycling safe and enjoyable. Bogotá has a network of hundreds of kilometers of bikeways under construction, and Santiago's Urban Transport Plan calls for building some 1,000 kilometers of bikeways.[15]

Done well, the bikeway strategy can be extremely successful, as experience in Europe demonstrates. The Netherlands has doubled the length of its network of bikeways in the past 20 years, and Germany has tripled its network.[16] Cycling accounts for some 12 percent of all trips in Germany, and for some 27 percent in the Netherlands, compared with less than 1 percent in the United States, where bicycle infrastructure is much less extensive and less sophisticated.[17] In addition, there are about four times as many cycling fatalities per kilometer traveled in the United States as in Germany or the Netherlands.[18]

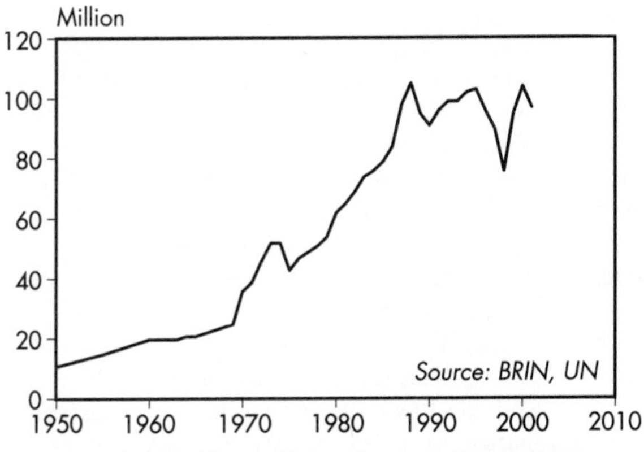

Figure 1: World Bicycle Production, 1950–2001

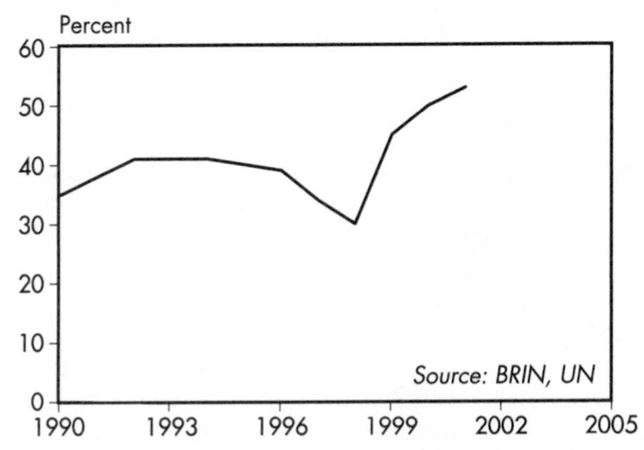

Figure 2: Chinese Bicycle Production as a Share of World Production, 1990–2001

World Bicycle Production, 1950–2001

Year	Production (million)
1950	11
1955	15
1960	20
1965	21
1970	36
1971	39
1972	46
1973	52
1974	52
1975	43
1976	47
1977	49
1978	51
1979	54
1980	62
1981	65
1982	69
1983	74
1984	76
1985	79
1986	84
1987	98
1988	105
1989	95
1990	91
1991	96
1992	99
1993	99
1994	102
1995	103
1996	96
1997	90
1998	76
1999	95
2000	104
2001	97

Source: *Bicycle Retailer and Industry News* and United Nations.

Communications Networks Expand

Molly O. Sheehan

Both mobile phones and the Internet attracted new users at double-digit rates in 2002, albeit more slowly than in the 1990s. The number of mobile or cellular phone subscribers worldwide in 2002 topped 1.15 billion, an increase of 21 percent over 2001.[1] (See Figure 1.) For the first time, mobile phones outnumbered fixed-line phone connections (1.05 billion).[2] At the same time, use of the Internet expanded, thanks in part to a 16.5-percent increase in host computers in 2002 to 171.6 million, drawing more than 600 million people online regularly.[3] (See Figure 2.)

LINKS p. 62

Within just one decade, the ranks of people communicating by wireless phones and wired computers have swelled significantly. In 1992, only one in 237 people worldwide used a mobile phone, and one in 778 used the Internet; by 2002, the numbers had soared to one in 5 and one in 10, respectively.[4] Today, well over 90 percent of all nations have local cell phone and Internet service, whereas in 1992, a person could use a cell phone in only one third of all countries and hook up to the Internet through a local number in just 19 percent.[5]

By linking computers with phones, the Internet sped the convergence of communications and computing technologies in the 1990s.[6] Now, as mobile phones proliferate, more people are making wireless connections to the Internet.[7] Since 2002, a growing number of U.S. cities have set up systems in parks and public spaces that give free Internet access to people with wireless modems in their laptop computers.[8] In Europe, more people send and receive short text messages with their cell phones than use the Internet from personal computers.[9]

Cell phones have helped bridge the telephonic divide between rich and poor. Building towers for them is cheaper than stringing copper wires for fixed-line phones, so start-up mobile services can recoup their investments and expand their coverage more quickly. As the average price of mobile phones has dropped by nearly 10 percent a year, it has fallen within reach of more people.[10] Between 1992 and 2001, phone penetration—the number of fixed lines and of mobiles per 100 people—accelerated in developing nations.[11] (See Figure 3.)

Gaps in phone access have closed more quickly in some countries than others, with striking differences among the nations in transition from planned to market economies. Central European nations, quick to invite mobile competition, saw dramatic gains in phone penetration in the 1990s; Hungary, for instance, went from 9.6 phone connections per 100 people in 1990 to 67.4 per 100 in 2000.[12] During the same period, in former Soviet republics in Central Asia, where the state still controls most telecom services, phone penetration did not grow beyond 20 per 100.[13]

Mobile service has dramatically increased access to phones in Africa. Uganda, the one nation where all three of Africa's leading cellular companies compete, in 1999 became the first country in that continent to have more mobile than fixed-line customers.[14] Some 30 other African nations have followed, as more people have hooked into the phone network in a few years of cellular expansion than in all the decades since independence.[15] Mobiles outnumber fixed lines in Africa today at a higher ratio than on any other continent.[16]

A greater gap separates those with and without Internet access, but this digital divide is also narrowing. In 2001, the industrial world had 41 Internet users per 100 people, whereas developing nations had 2.3 per 100—still a 17 to 1 ratio, but much better than the 40 to 1 ratio in 1995.[17] By linking rural farmers to market information, craftworkers to customers, patients to doctors, and students to teachers, the Internet can aid economic development.

Cheap computers with nonproprietary software, designed to be shared at public libraries, cyber cafes, and telecenters, could bring the Internet to even more people. Indian scientists have built a handheld "Simputer"—short for Simple, Inexpensive, Multilingual Computer—for poor, rural users that is expected to reach the market in 2003.[18] While the computer itself will cost about $200, people will be able to rent time on one—for instance, to check commodity prices or consult doctors—and to store their own data on $1–2 cards.[19]

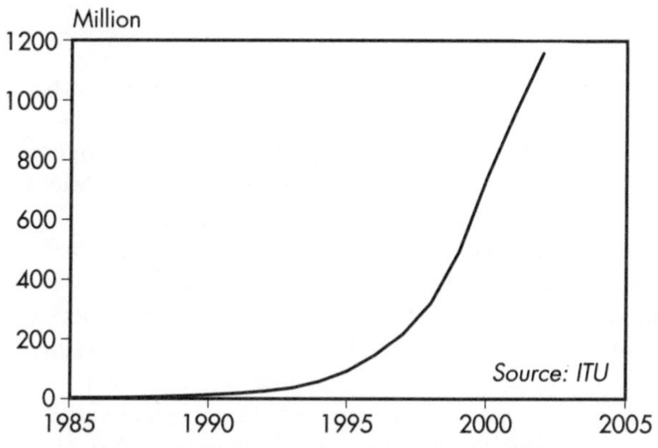

Figure 1: Cellular Phone Subscribers Worldwide, 1985–2002

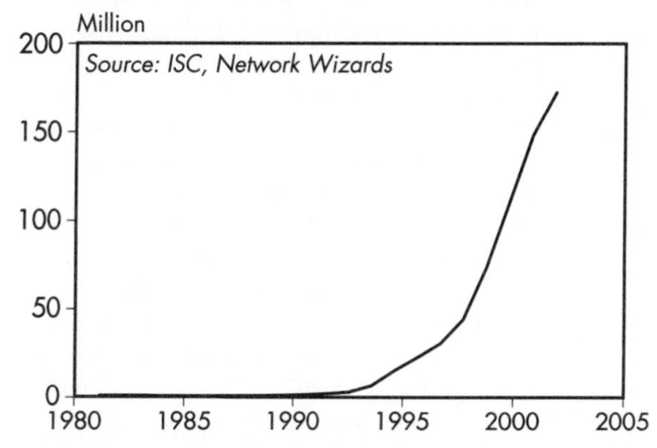

Figure 2: Internet Host Computers, 1981–2002

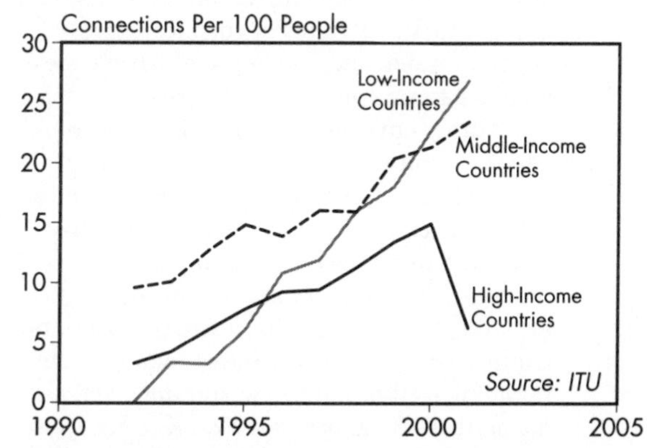

Figure 3: Annual Growth Rate in Phone Connections (Cellular and Fixed-Line) by Income Level of Country, 1992–2001

Cellular Phone Subscribers and Internet Host Computers Worldwide, 1985–2002

Year	Cellular Phone Subscribers	Internet Host Computers
	(million)	(number)
1985	1	2,308
1986	1	5,089
1987	2	28,174
1988	4	80,000
1989	7	159,000
1990	11	376,000
1991	16	727,000
1992	23	1,313,000
1993	34	2,170,000
1994	56	5,846,000
1995	91	14,352,000
1996	144	21,819,000
1997	215	29,670,000
1998	319	43,230,000
1999	491	72,398,092
2000	741	109,574,429
2001	955	147,344,723
2002	1,155	171,638,297

Source: International Telecommunication Union, Internet Software Consortium, and Network Wizards.

Semiconductor Sales Rebound Slightly

David Taylor

Sales of semiconductors, the brains behind modern electronics, rose in 2002 after the largest crash in nearly 50 years of semiconductor manufacturing.[1] (See Figure 1.) During 2000, semiconductor sales surged from $150 billion to $209 billion (in 2001 dollars), setting the record for the largest one-year rise in sales and the strongest sales ever.[2] The following year provided the antithesis: the largest one-year drop in history to a level just below that in 1996—$139 billion. Sales remained virtually stagnant in 2002, with a gain of only $500 million.[3]

On a decadal time scale, however, the semiconductor industry is growing. From 1970 to 2002, the average annual growth for the industry was 9 percent, and during the 1980s and 1990s the industry grew 1.8 and then 2.8 times faster than the global economy.[4] Despite continued demand for its products, the industry has collapsed four times in its history: a crash of $7.6 billion in 1985, of $16.6 billion in 1996, of $14.1 billion in 1998 during the Asian financial crisis, and of $70.3 billion in 2001.[5]

LINKS pp. 44, 60

A combination of factors led to the recent losses, most notably the dot-com bust in early 2000 and a global recession that picked up steam mid-2001.[6] Some analysts also blame an inventory glut spurred by poor forecasting, with semiconductor (chip) manufacturers overproducing to gain nonexistent market share.[7]

The 2002 growth was led solely by the Asia-Pacific region (excluding Japan), where the market grew by 31 percent.[8] (See Figure 2.) China continues to be a focus for chip makers.[9] The International Expo Group forecasts the Chinese market to reach almost 9 percent of world semiconductor sales in 2003, making it the second-largest market in the world.[10] Sales in the Americas, Europe, and Japan all continued to decline in 2002.[11]

Although chips are bought around the world, production is primarily confined to East Asia, Europe, and North America. A 1999 survey of U.S.-based companies found that 59 percent of their employees were located in the United States, 33 percent in Asia, 6 percent in Europe, and 2 percent in Latin America.[12]

To avoid the cost and liability of a $3–5 billion fabrication facility, most semiconductor firms outsource some or all of their production to the "Big Three" foundries—Taiwan Semiconductor Manufacturing Company, United Microelectronics Corporation in Taiwan, and Chartered in Singapore.[13] Major firms are currently outsourcing fewer products, fighting to keep their own plants at full capacity, although AMD and Motorola indicate that in a market upswing they plan to outsource 25 percent and 50 percent of their production, respectively.[14]

Semiconductors undoubtedly have positive environmental results for society, such as the ability to telecommute and the heightened efficiency of industrial processes. Yet issues surrounding production and disposal, as well as the health of production workers, cloud this industry's sleek image.[15]

According to a recent study, the negative environmental impacts of high-tech are manifest to a greater degree in the production phase of a chip's life-cycle; this contrasts considerably with a consumer product like an automobile, where the major environmental impacts arise in the use phase.[16] The total weight of fossil fuels and chemicals used to produce a 2-gram DRAM, or memory chip, is 630 times the weight of the chip itself, pointing to the existence of a secondary materials stream.[17] Over its life cycle, which includes both production and use, this 2-gram chip requires 1,600 grams of fossil fuels, 72 grams of chemicals, 32,000 grams of water, and 700 grams of harmless elemental gases during its life.[18]

Another environmental problem is disposal, which is compounded by the complexity of the devices and their short-term usefulness. In the past, an average computer lasted four to five years; the trend today is retirement after just two years.[19] One study predicts greater recycling of this electronic waste (e-waste), but another study recently exposed toxic e-waste recycling facilities in China and India and estimated that 50–80 percent of e-waste generated in the United States is exported for recycling.[20]

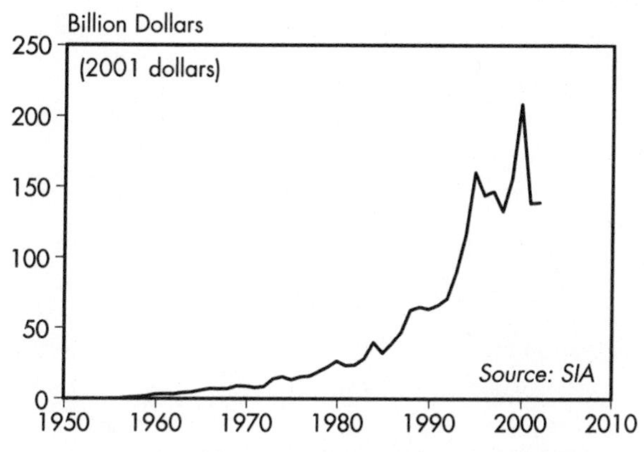

Figure 1: World Semiconductor Sales, 1954–2002

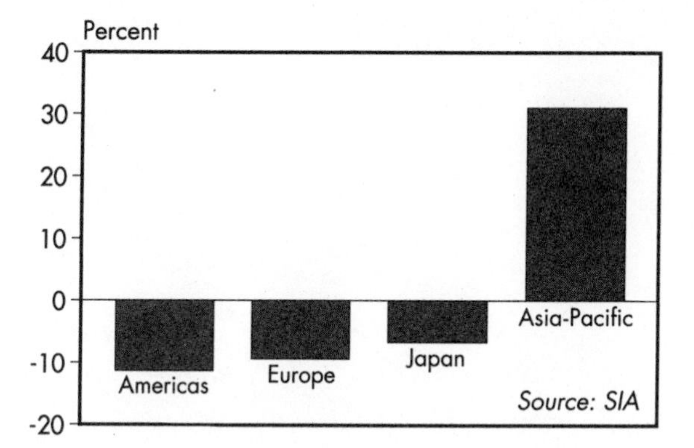

Figure 2: Regional Breakdown of Semiconductor Sales, 2002

World Semiconductor Sales, 1954–2002

Year	Sales (billion 2001 dollars)
1954	0.03
1960	3.20
1965	5.98
1970	8.85
1971	7.89
1972	8.60
1973	14.00
1974	15.54
1975	13.39
1976	15.52
1977	16.04
1978	19.28
1979	22.41
1980	26.66
1981	23.51
1982	23.95
1983	28.44
1984	39.98
1985	32.37
1986	39.23
1987	47.11
1988	63.16
1989	65.31
1990	63.87
1991	66.63
1992	71.37
1993	89.93
1994	116.13
1995	161.06
1996	144.43
1997	147.25
1998	133.17
1999	156.67
2000	209.24
2001	138.90
2002 (prel)	139.40

Source: Semiconductor Industry Association.

Health and Social Trends

JHU/CCP

Population Growth Slows

HIV/AIDS Pandemic Spreads Further

Cigarette Production Dips Slightly

Population Growth Slows

Molly O. Sheehan

Between 2001 and 2002, births exceeded deaths by 74 million, pushing world population over 6.2 billion.[1] (See Figure 1.) Last year's growth of 1.18 percent was the lowest since rates peaked above 2 percent in the mid-1960s.[2] With a larger population growing at a slower rate, the number of people added to the planet annually has in fact remained about the same, and the human family has more than doubled since 1960.[3]

Although deaths from AIDS and lower than expected fertility prompted the United Nations to reduce its global population projections—to 8.9 billion people by 2050, not 9.3 billion—the 49 poorest countries in the world still have populations that are increasing at 2.4 percent per year, nearly 10 times the 0.25 percent annual growth in industrial nations.[4] All the countries with the highest birth rates are among the world's poorest. (See Figure 2.)

LINKS pp. 106, 108

Population growth is slowest in nations that moved from Communist rule in the 1990s, as higher mortality and higher emigration followed the collapse of economies. A growing gap in life expectancy divides Western Europe, where only 10 percent live below the poverty line, from Central and Eastern Europe and the former Soviet Union, where the share of the people living on less than $4 a day skyrocketed from 3.3 percent in 1988 to 46 percent—nearly half the populace—by the end of the 1990s.[5]

Sharp declines in birth rates in a few populous nations are largely responsible for the slower growth of world numbers since the late 1960s. In Indonesia, the average number of children born to each woman in 1950–55 was 5.5; by 1995–2000, the figure had fallen to 2.6.[6] Over the same period, fertility fell in Brazil from 6.2 to 2.3 children.[7] By 2000, fertility had dropped below the replacement level of 2.1 children per woman in 17 nations in the developing world, including China.[8]

Lower fertility and slower population growth have been linked, since 1970, with economic development in Brazil, Mexico, and several East Asian nations.[9] With better health care, death rates declined. And with better access to contraceptives, people had fewer children and more women could work outside the home. A demographic window of opportunity for development opened, with a large group of working-age people supporting relatively fewer dependents—both older and younger.[10]

The labor force in Indonesia, Singapore, South Korea, Taiwan, and Thailand grew more rapidly than total population by about 25 percent a year between 1960 and 1990—a demographic bonus that boosted per capita income by 0.8 percent a year.[11] Similarly, declining fertility in Brazil increased per capita income by 0.7 percent annually.[12] With a smaller share of the population in school, countries could raise spending per child. Analysts credit the East Asian "economic miracle" to public investments made in education, health care, and opportunities for women.[13]

Population growth is increasingly concentrated in cities. As the pace of growth in cities has outstripped that of rural areas for more than a century, the share of the world living in urban areas has grown steadily—from 10 percent in 1900 to 30 percent in 1950 to nearly 48 percent in 2001.[14] The United Nations estimates that some 2 billion people will be added to world population between 2000 and 2030—nearly all of them in urban areas of the developing world.[15] In this "medium-growth" scenario, the urban share of world population will pass 50 percent in 2007 and top 60 percent by 2030.[16] (See Figure 3.)

As population swells in urban centers of poor nations and wanes in some richer nations, more people are likely to migrate. The United States is the only industrial nation with a fertility rate still above replacement level, partly due to immigrants.[17] In Japan, whose population is aging faster than any other nation's, the average age is expected to rise from 41 to 53 between 2000 and 2050 as the population contracts by 14 percent.[18] Populations are also expected to age and shrink throughout Europe, where concern about the strain on pension and health care systems is mounting at the same time that rising numbers of migrants face an anti-immigrant backlash.[19]

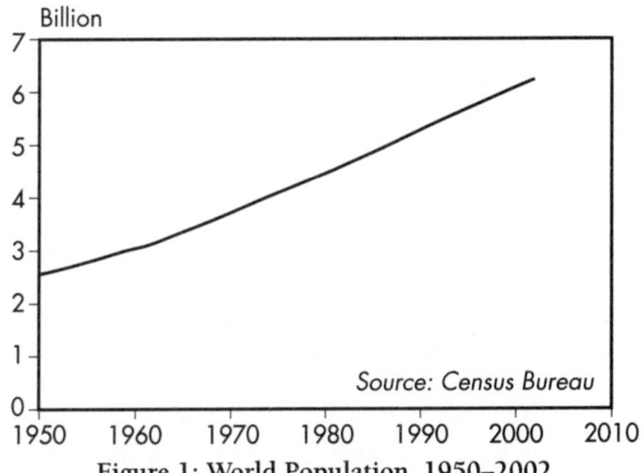

Figure 1: World Population, 1950–2002

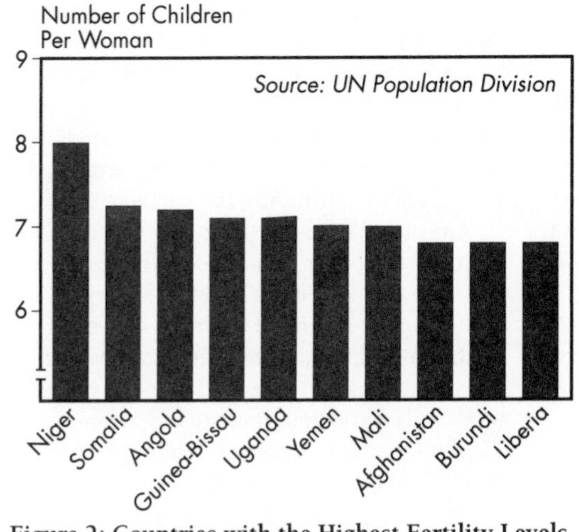

Figure 2: Countries with the Highest Fertility Levels, 2000–05

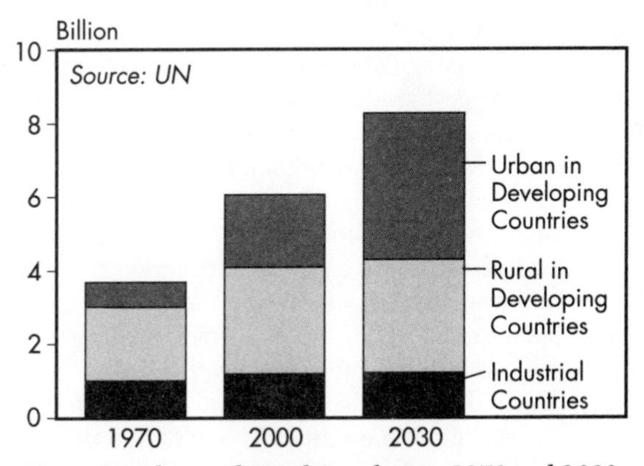

Figure 3: Urban and Rural Population, 1970 and 2000, with Projections for 2030

World Population, Total and Annual Addition, 1950–2002

Year	Total[1] (billion)	Annual Addition (million)
1950	2.555	38
1955	2.780	53
1960	3.040	41
1965	3.346	70
1970	3.708	78
1971	3.786	77
1972	3.862	76
1973	3.939	76
1974	4.014	73
1975	4.087	72
1976	4.159	72
1977	4.232	72
1978	4.304	75
1979	4.379	76
1980	4.455	76
1981	4.530	80
1982	4.611	80
1983	4.691	79
1984	4.770	80
1985	4.850	82
1986	4.932	85
1987	5.017	86
1988	5.103	86
1989	5.189	87
1990	5.275	84
1991	5.359	84
1992	5.443	81
1993	5.524	80
1994	5.605	81
1995	5.685	79
1996	5.764	80
1997	5.844	79
1998	5.923	78
1999	6.002	77
2000	6.079	75
2001	6.154	74
2002 (prel)	6.228	74

[1] Total at mid-year.
Source: U.S. Bureau of the Census.

HIV/AIDS Pandemic Spreads Further

Radhika Sarin

The number of people living with HIV/AIDS rose to 42 million at the end of 2002.[1] Five million people became infected with HIV in 2002 (see Figure 1), and another 3.1 million died of AIDS-related causes.[2] (See Figure 2.)

For the first time, women account for half the people living with HIV/AIDS.[3] Heterosexual transmission, particularly in Africa and the Caribbean, is the primary cause of infection among women, who are two to four times more likely than men to become infected during unprotected vaginal sex.[4]

Women's biological vulnerability—due to a large surface area of reproductive tissue and high virus concentrations in infected semen—is compounded by economic and social inequities. Women who are economically dependent on husbands or sexual partners have little control over sexual relations and condom use. Social taboos prevent women from learning about reproductive health, while the stigma associated with sexually transmitted infections is a barrier against seeking care.[5] Young women risk contracting HIV from older partners, which explains large differences in HIV prevalence between teenaged males and females in many countries.[6]

LINKS pp. 104, 108, 110

In sub-Saharan Africa, home to 70 percent of the world's HIV-positive people, AIDS is the leading cause of death.[7] In 2002, average life expectancy in 16 African nations was at least 10 years lower than it would have been without AIDS.[8] HIV/AIDS is also exacerbating Africa's food crisis, threatening about 38 million people with starvation.[9] The epidemic has reduced the number of agricultural workers and has unraveled social safety nets as families sell off assets to pay for medical or funeral expenses. For the poor without access to antiretroviral therapies, good nutrition is all that can ward off illness and early death. When this is also lost, weakened immune systems become susceptible to tuberculosis, malaria, and other infections.[10]

Though Africa carries the greatest burden of disease, the epidemic is growing fastest in Eastern Europe and Central Asia, where it is linked to intravenous drug use, high unemployment, and crumbling public health facilities. In Russia, up to 90 percent of registered infections are due to drug use.[11] High rates of sexually transmitted infections in the region indicate that heterosexual transmission could spread HIV into the wider population, as seen in Ukraine and Belarus.[12] In the Baltic states, overcrowded prisons and juvenile justice institutions serve as breeding grounds for the virus.[13]

Another emerging AIDS hotspot is Asia, where low national prevalence levels in populous nations mask the magnitude of localized infection. Nearly 4 million people are infected in India, and the epidemic has hit the general population in several states.[14] In China, people in poor rural communities who participated in blood-selling programs in the 1990s have become concentrated pockets of HIV-positive villagers with limited access to any kind of care.[15] In all, China reports an estimated 1 million infections, with drug use and heterosexual transmission continuing the spread.[16]

Although AIDS-related mortality has fallen dramatically in high-income countries since antiretroviral treatment became widespread in 1996, only 4 percent of those who need treatment in low- and middle-income countries receive it.[17] (See Figure 3.) The price of antiretrovirals has fallen dramatically, from $10,000–12,000 a year per person in early 2000 to $350 by December 2001.[18] The world's poorest, however, cannot afford even this.

Yet some progress has been made in making access to treatment more equitable. In 2002, Botswana became the first African nation to adopt a policy of universal access to treatment.[19] The biggest strides have been made in Latin America and the Caribbean: Brazil, Argentina, Costa Rica, Cuba, Uruguay, Honduras, and Panama are among the countries providing free or subsidized treatment.[20]

Even so, a huge gap persists between needed and available resources. In 2002, $3 billion was spent on efforts to stem the epidemic in low- and middle-income countries.[21] UNAIDS estimates that funding will have to more than double, to $6.5 billion, in 2003—with two thirds of the resources coming from international sources.[22]

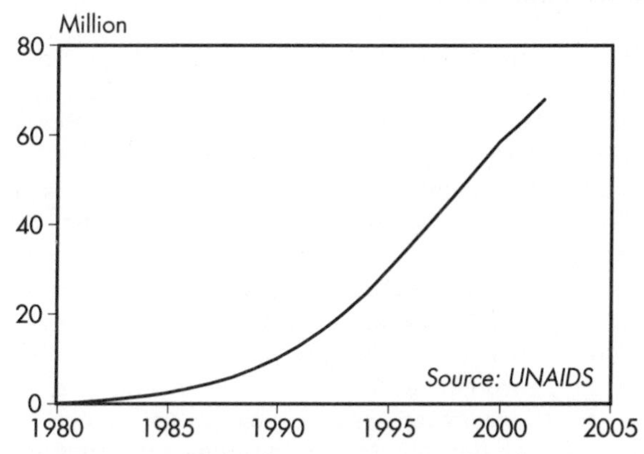

Figure 1: Estimates of Cumulative HIV Infections Worldwide, 1980–2002

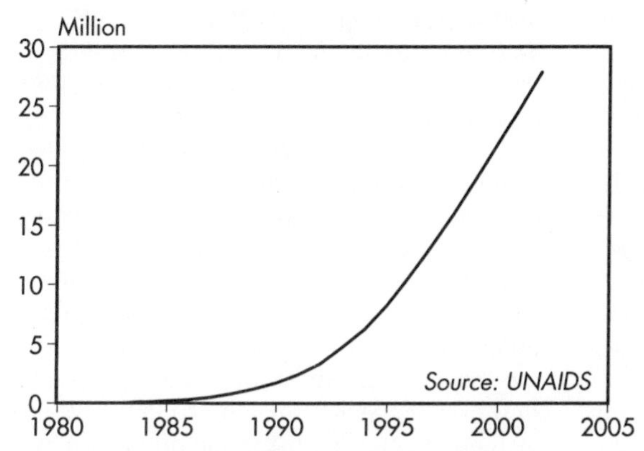

Figure 2: Estimates of Cumulative AIDS Deaths Worldwide, 1980–2002

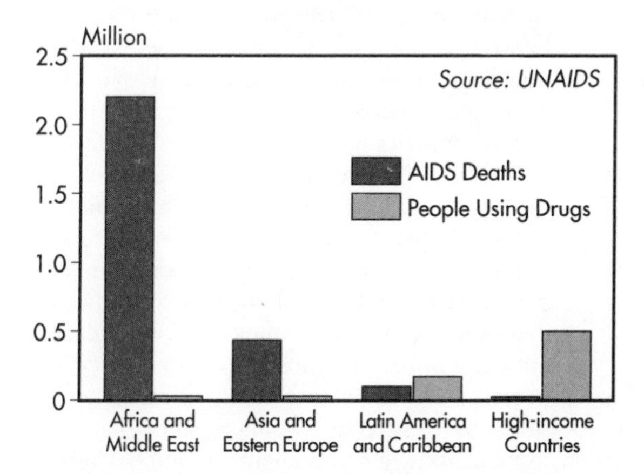

Figure 3: AIDS Deaths and Number of People Using Antiretroviral Drugs, by Region, 2001

Cumulative HIV Infections and AIDS Deaths Worldwide, 1980–2002

Year	HIV Infections	AIDS Deaths
	(million)	
1980	0.1	0.0
1981	0.3	0.0
1982	0.7	0.0
1983	1.2	0.0
1984	1.7	0.1
1985	2.4	0.2
1986	3.4	0.3
1987	4.5	0.5
1988	5.9	0.8
1989	7.8	1.2
1990	10.0	1.7
1991	12.8	2.4
1992	16.1	3.3
1993	20.1	4.7
1994	24.5	6.2
1995	29.8	8.2
1996	35.3	10.6
1997	40.9	13.2
1998	46.6	15.9
1999	52.6	18.8
2000	58.5	21.8
2001	62.9	24.8
2002 (prel)	67.9	27.9

Source: Joint United Nations Programmme on HIV/AIDS.

Cigarette Production Dips Slightly

Erik Assadourian

Global cigarette production fell to 5.6 trillion pieces in 2002, a decrease of 0.5 percent over 2001.[1] (See Figure 1.) While total production has hovered around the same mark for the past decade, population growth during this time has reduced per capita output 13 percent since 1990, to 897 cigarettes per person a year.[2] (See Figure 2.)

China, the United States, and Russia—the three largest producers—manufacture just under half of the world's supply. In 2002, China produced 1.7 trillion cigarettes, 31 percent of global production.[3] The United States manufactured 580 billion cigarettes, or 10 percent.[4] But unlike China, which uses 99 percent of the cigarettes it produces, the United States exports 23 percent of its output.[5]

LINKS pp. 48, 108

Russia, traditionally a smaller producer, is now the third largest, manufacturing 375 billion cigarettes in 2002, more than twice as many as in 1998.[6] And Russians are now leaders in per capita cigarette consumption—smoking 1,931 cigarettes in 2002, more than twice the global average.[7] (See Figure 3.)

Of the more than 1.1 billion smokers worldwide, 82 percent live in low- or middle-income countries.[8] Between high population growth and aggressive tobacco marketing campaigns in these regions, most of the growth in smoking is expected to occur in these nations—a development that will increasingly burden public health systems already straining from a lack of resources and from diseases like AIDS.[9]

Currently, smoking kills 4.9 million people a year—one in 10 adult deaths—from a range of illnesses that includes heart disease, various forms of cancer, and stroke.[10] By 2030, experts foresee smoking becoming the leading cause of death, responsible for 10 million deaths a year—of which 7 of every 10 would occur in low- or middle-income countries.[11]

Globally, cigarettes and cigarette lights (matches and lighters) also cause 17,000 fire deaths and $27 billion of damage each year.[12] In the United States, cigarettes cost $76 billion a year in health care expenditures and another $82 billion in lost productivity.[13] Secondhand smoke also threatens health, increasing the risk of lung cancer and heart disease more than 20 percent.[14]

Since 1999, a coalition led by the World Health Organization has been drafting a Framework Convention on Tobacco Control to reduce consumption through measures that include stronger labeling requirements, marketing restrictions, anti-smuggling laws, and workplace bans.[15] This global treaty will be ready for signature in 2003, but its success is uncertain, as the tobacco industry and several governments have tried to weaken the text.[16]

Many regions have already significantly reduced smoking by controlling tobacco. Cigarette taxes lower smoking rates while providing governments with funds to combat smoking-related health problems.[17] In the United Kingdom, as cigarette prices increased in real terms by 70 percent over two decades, consumption declined by more than 35 percent.[18]

Counter-advertising, such as anti-smoking commercials and explicit health labels placed on cigarette packs, also helps reduce smoking.[19] In Canada, a 2001 survey showed that 90 percent of smokers noticed the visually disturbing labels and 44 percent were more motivated to quit.[20] Such efforts alone cannot combat the huge marketing budgets of tobacco companies. In the United States, the tobacco industry spent $9.6 billion on advertising and promotion in 2000.[21] Restrictions have little effect, as the industry just shifts to new marketing mediums, such as sponsoring sporting events.[22] Comprehensive marketing bans are more successful, however—decreasing smoking by up to 6.3 percent over two decades.[23] Currently, more than a dozen countries have such bans.[24]

In 2002, Thailand banned smoking in indoor public spaces, the strictest smoking ban in Asia.[25] Smoking bans have proved successful in curbing smoking and reducing exposure to secondhand smoke.[26] A review of 26 studies determined that totally smoke-free workplaces cut the number of cigarettes consumed by 29 percent—the equivalent of raising cigarette prices by 73 percent.[27] Canada, since implementing a comprehensive national anti-tobacco strategy in 1999, has cut the national smoking rate by 3 percent.[28]

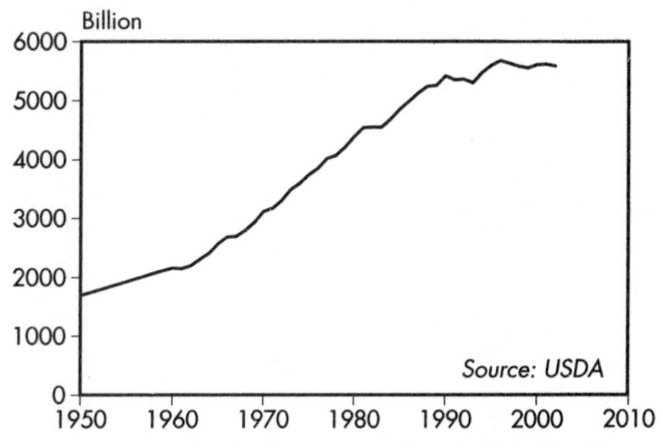

Figure 1: World Cigarette Production, 1950–2002

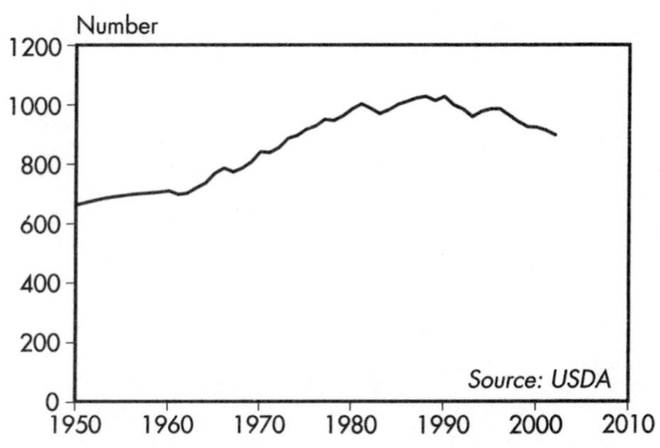

Figure 2: World Cigarette Production Per Person, 1950–2002

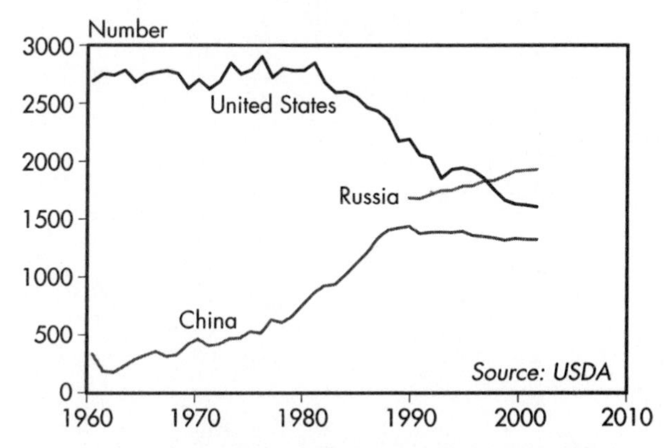

Figure 3: Cigarette Consumption Per Person in the United States and China, 1960–2002, and in Russia, 1990–2002

World Cigarette Production, 1960–2002

Year	Total (billion)	Per Person (number)
1950	1,686	660
1955	1,921	691
1960	2,150	707
1965	2,564	766
1970	3,112	839
1971	3,165	836
1972	3,295	853
1973	3,481	884
1974	3,590	894
1975	3,742	915
1976	3,852	926
1977	4,019	950
1978	4,072	946
1979	4,214	962
1980	4,388	985
1981	4,541	1,002
1982	4,550	987
1983	4,547	969
1984	4,689	983
1985	4,855	1,001
1986	4,987	1,011
1987	5,128	1,022
1988	5,240	1,027
1989	5,258	1,013
1990	5,419	1,027
1991	5,351	998
1992	5,363	985
1993	5,300	959
1994	5,478	977
1995	5,599	985
1996	5,680	985
1997	5,633	964
1998	5,581	942
1999	5,554	925
2000	5,609	923
2001	5,617	913
2002 (prel)	5,587	897

Source: U.S. Department of Agriculture; data for 1950–59 are estimates based on USDA data.

Military Trends

UN/DPI Photo

Violent Conflicts Continue to Decline

Peacekeeping Expenditures Down Slightly

Violent Conflicts Continue to Decline

Michael Renner

According to AKUF, a conflict research group at the University of Hamburg, the number of wars worldwide stood at 28 in 2002, down from 31 the previous year.[1] In addition, there were 17 "armed conflicts" active in 2002 that were not of sufficient severity to meet AKUF's criteria for war.[2] Combining both categories, the total number of violent clashes declined slightly—from 48 in 2001 to 45.[3] (See Figure 1.)

The overall number declined because the number of conflicts ending—those in the Kurdish areas of eastern Turkey, the Democratic Republic of the Congo, Guinea, Kosovo, Iran, Tajikistan, and Uzbekistan—surpassed those newly erupting—in Côte d'Ivoire, Madagascar, Congo-Brazzaville, and the Central African Republic.[4]

LINKS pp. 76, 102, 118, 120

Meanwhile, the U.S.-led "war on terror," initially focused on the Afghan Taliban regime, more and more has the makings of an open-ended campaign of worldwide scope. The Bush administration's words and actions made it seem all but inevitable that an invasion of Iraq would occur in 2003.[5] Other countries, including Russia, China, India, Indonesia, and Israel, have also cited anti-terrorism as an excuse for wars or acts of internal repression.[6]

The armed forces of countries on whose territory fighting is taking place number in the millions, but it is unclear how many of their soldiers are actually engaged in combat. Non-state armed groups worldwide have at least some 350,000 fighters.[7] Of these, about 140,000 were with groups that observed cease-fires or were otherwise inactive in 2002.[8] Some 300,000 children are among government or opposition forces involved in fighting.[9]

Measuring whether the world is becoming more or less violent is not an easy task.[10] Information is often incomplete or contradictory. And definitional and methodological problems confound efforts to establish unambiguous categories and thresholds to tally the number of armed conflicts. Findings from different research groups, therefore, offer some variations in their findings—reflecting a complex reality. Figures compiled by researchers in Sweden and Norway, for instance, show a substantial number of unclear cases.[11] (See Figure 2.)

Researchers at the Heidelberg Institute for International Conflict Research in Germany are assessing political conflict trends from a broader perspective. The total number of conflicts has increased from 108 in 1992 to 173 in 2002.[12] Of these, violent conflicts have recently accounted for a fairly steady one-quarter share.[13] (See Figure 3.) The 17 conflicts that escalated during 2002 were more than outweighed by 31 de-escalated cases.[14]

The majority of conflicts are resolved by nonviolent means, including negotiations and other diplomatic efforts. In addition to various peacekeeping efforts, negotiations were taking place in 43 conflicts in 2002, resulting in three peace treaties (in Chad, Moluccas, and Aceh) and seven cease-fire agreements (which were successful in Angola, Sri Lanka, and Somalia).[15] U.N. arms embargoes and other sanctions were maintained in eight cases.[16]

Researchers at PIOOM in the Netherlands have made an even more extensive effort to capture a broader multitude of conflicts, including intercommunal conflicts not recorded elsewhere. PIOOM finds that there are more than 300 "political tension situations"—hard-to-monitor cases that typically either predate violent conflict or follow it, possibly giving rise to renewed violence.[17] These findings underscore that today's human rights violations, inequalities, and environmental destruction often end up as tomorrow's wars.

The number of wars alone cannot of course convey the severity of warfare in terms of human suffering, political instability, or social, economic, and environmental damage inflicted. AKUF estimates that more than 7 million people, most of them civilians, have died in the course of the 45 wars and armed conflicts currently active.[18] Cumulatively, these conflicts have cost at least $250 billion—imposing a heavy toll on countries that for the most part are already desperately poor.[19] And the cost of reconstruction could be much higher.

Figure 1: Wars and Armed Conflicts, 1950–2002

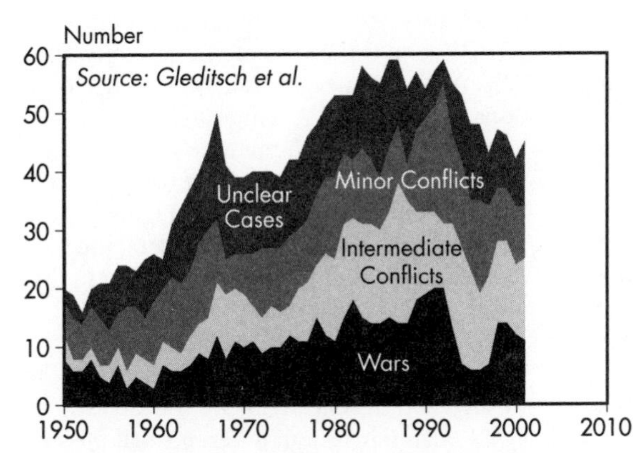

Figure 2: Wars, Intermediate and Minor Conflicts, and Unclear Cases, 1950–2001

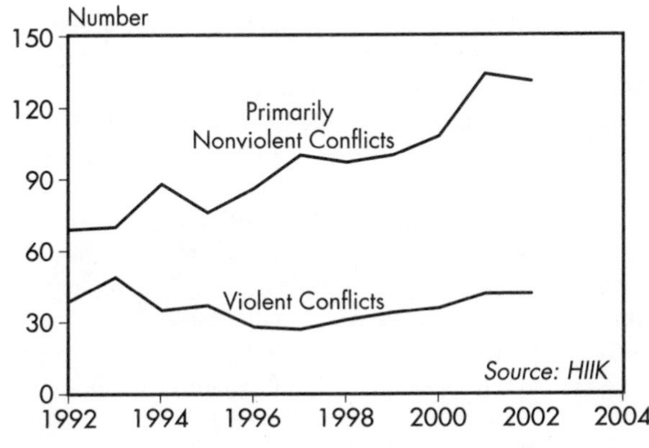

Figure 3: Violent and Nonviolent Conflicts, 1992–2002

Wars and Armed Conflicts, 1950–2002

Year	Wars	Wars and Armed Conflicts
	(number)	
1950	13	
1955	15	
1960	12	
1965	28	
1970	31	
1971	31	
1972	30	
1973	30	
1974	30	
1975	36	
1976	34	
1977	36	
1978	37	
1979	38	
1980	37	
1981	38	
1982	40	
1983	40	
1984	41	
1985	41	
1986	43	
1987	44	
1988	45	
1989	43	
1990	50	
1991	54	
1992	55	
1993	48	65
1994	44	61
1995	34	49
1996	30	48
1997	29	47
1998	33	50
1999	35	49
2000	35	47
2001	31	48
2002 (prel)	28	45

Source: Arbeitsgemeinschaft Kriegsursachenforschung and the Institute for Political Science at the University of Hamburg.

Peacekeeping Expenditures Down Slightly

Michael Renner

Expenditures for United Nations peacekeeping operations from July 2002 to June 2003 are expected to reach about $2.6 billion—slightly less than in the previous reporting period.[1] (See Figure 1.) This contrasts with military expenditures worldwide of $839 billion in 2001.[2] Some 47,000 soldiers, military observers, and civilian police served in peacekeeping missions during 2002, but the number dropped to below 40,000 by the end of the year.[3] (See Figure 2.) In addition, the missions were aided by 10,929 civilians.[4] Two missions—in Bosnia and Croatia—ended in December 2002.[5]

LINKS pp. 76, 102, 118, 120

In addition to peacekeeping operations with strong military and police components, the United Nations also maintained 13 small "political and peace-building" missions during 2002, involving 1,073 mostly civilian staff.[6] The largest of these is the U.N. Assistance Mission in Afghanistan set up in March 2002.[7]

The permanent members of the Security Council have been reluctant to make significant troop commitments to peacekeeping missions. The leading contributors of personnel are Bangladesh, Pakistan, Nigeria, India, and Ghana, together accounting for 43 percent of the total.[8]

By far the largest current operation, with more than 16,000 peacekeepers, is in Sierra Leone, where a gruesome civil war fueled by diamond wealth has now wound down.[9] In December 2002, the Security Council decided to raise the authorized personnel strength of the mission in the Democratic Republic of the Congo from 4,250 to 8,700.[10] Following the withdrawal of foreign armies, there is hope that a December 2002 peace accord will end violence among domestic combatants in eastern Congo.[11]

In four other locations, the United Nations maintains missions that each deploy 3,000–5,500 peacekeepers.[12] In addition to Kosovo, the Ethiopia-Eritrea border, and southern Lebanon, this includes the U.N. Mission in Support of East Timor, a follow-on to the transitional administration that facilitated East Timor's May 2002 independence.[13]

The Sierra Leone mission costs about $700 million a year, followed by the Congo operation at $608 million.[14] Expenditures for the Kosovo and East Timor deployments run to more than $300 million each, and the Ethiopia-Eritrea observers cost more than $200 million.[15]

As of 31 December 2002, U.N. members still owed the organization $1.34 billion for peacekeeping operations.[16] The United States accounted for 40 percent of unpaid dues, or $536 million.[17] The next-largest amounts were owed by Japan ($312 million), Italy ($41 million), China ($39 million), Spain ($32 million), and Brazil ($28 million).[18]

A substantial number of peacekeeping missions are also being carried out by regional organizations such as NATO, the Organization for Security and Co-operation in Europe (OSCE), and the Economic Community of West African States, as well as by ad hoc coalitions of states.[19] In recent years, there have been 30–40 non-U.N. missions, involving a far larger number of peacekeeping troops than the United Nations deploys.[20] (See Figure 3.) Although information is incomplete, the combined cost of these operations is likely in the range of $8–12 billion a year.[21]

NATO-led deployments in the Balkans, where more than 50,000 soldiers are patrolling Kosovo and Bosnia, are by far the largest.[22] The OSCE maintains about a dozen small missions in Eastern Europe and former Soviet republics, involving about 1,500 people.[23] A multinational observer force of roughly 1,900 soldiers has been deployed since 1982 in the Sinai Peninsula.[24] Russia keeps some 5,000 troops in Moldova and the Caucasus.[25] And in Afghanistan, the International Security Assistance Force was created in December 2001 to ensure security in Kabul.[26] It has about 4,800 soldiers from 28 countries.[27] Governments have shortsightedly rejected suggestions that this force be extended beyond Kabul, even though much of Afghanistan is again falling under the sway of warlords and lawlessness.[28]

The personnel devoted to all forms of peacekeeping—some 110,000 persons in 2002—is dwarfed by the more than 400,000 soldiers deployed abroad for traditional military purposes, more than half of whom are U.S. troops.[29]

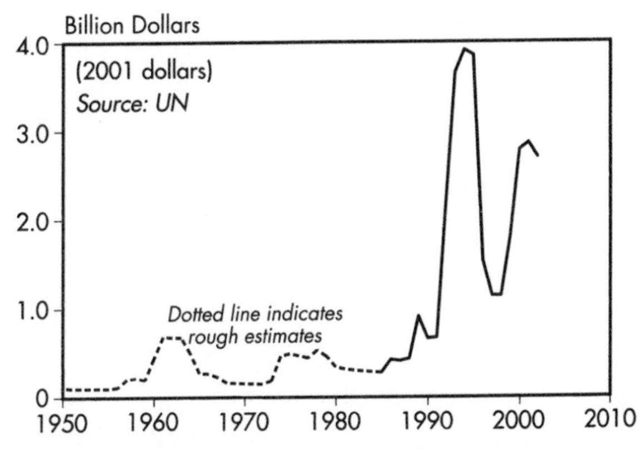

Figure 1: U.N. Peacekeeping Expenditures, 1950–2002

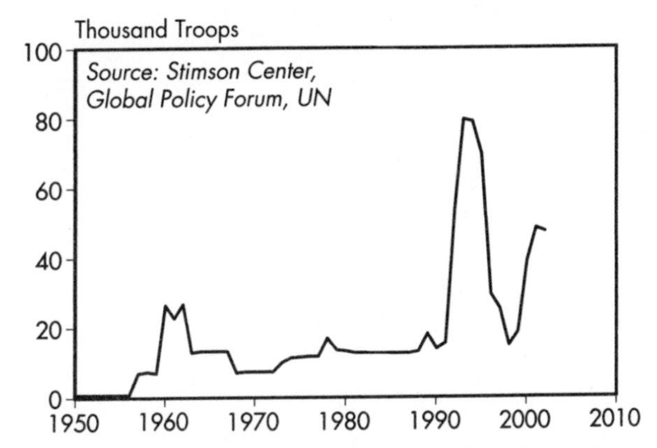

Figure 2: U.N. Peacekeeping Personnel, 1950–2002

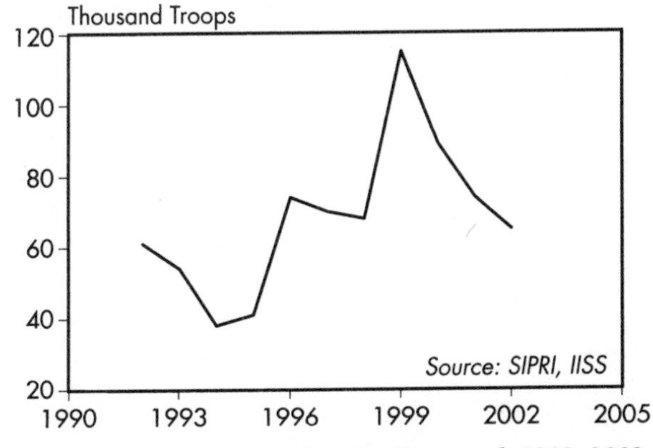

Figure 3: Non-U.N. Peacekeeping Personnel, 1992–2002

U.N. Peacekeeping Expenditures, 1986–2002

Year	Expenditure
	(billion 2001 dollars)
1986	0.352
1987	0.339
1988	0.363
1989	0.834
1990	0.587
1991	0.598
1992	2.105
1993	3.559
1994	3.809
1995	3.752
1996*	1.456
1997*	1.063
1998*	1.060
1999*	1.721
2000*	2.692
2001*	2.770
2002*	2.609

* July to June of following year.

Source: U.N. Department of Public Information and U.N. Department of Peacekeeping Operations.

Part Two

SPECIAL FEATURES

Environment Features

© Digital Vision

Birds in Decline

Small Islands Threatened by Sea Level Rise

Birds in Decline

Howard Youth

Around the world, ornithologists are alarmed at bird population declines and are concerned about what they mean for the world's ecosystems and our own future. In 2000, a study published by a global alliance of conservation groups called BirdLife International found that about 12 percent of the world's 9,800 bird species are threatened with extinction within the next century and that in the near future an additional 8 percent may become threatened.[1] In addition, the populations of many widely distributed and still plentiful species are in decline. (See Table 1.)

LINKS p. 40

These trends worry scientists because birds provide critical natural services—dispersing seeds, pollinating flowers, controlling insect and rodent populations, and scavenging dead animals.[2] Of course, birds' colorful plumage, songs, and varied behaviors also capture people's attention. To many, the world would seem incomplete without ostriches, eagles, flamingos, parrots, hummingbirds, and orioles.

In addition, many declining bird species serve as indicators of impending environmental problems. Aquatic songbirds called dippers, for example, disappear from stream waters acidified by pine plantations and acid rain.[3] Dying North American crows, hawks, and owls mark the rapid spread of the West Nile virus that has been introduced to the region.[4]

Since 1800, 103 species have gone extinct.[5] BirdLife International's threatened list tallies almost 1,200 more that may vanish within the next century.[6] Many are declining due to combined human-related threats. Habitat loss or degradation is now the single greatest overall threat to birds, including 85 percent of the most imperiled species. Many of these species are endangered by recent tropical forest destruction.[7]

In addition, many bird populations shrink after invasions by introduced, or exotic, species, which constitute the second greatest threat.[8] Rats, cats, mongooses, and other exotic predators kill birds and their young; introduced birds compete with native species; and exotic pathogens knock out endemic birds that lack disease resistance. Introduced insect pests destroy birds' forest habitats. Worldwide, exotic plants also alter local flora, denying birds their food supplies and habitat.[9]

Control of exotic species often requires costly active management that may include pesticides and other tools that might harm native fauna as well.[10] In the United States alone, estimates of the annual cost of damage caused by exotics and the measures to control them reach as high as $137 billion.[11]

Meanwhile, poorly regulated or illegal hunting and capture take a heavy toll. In Malta, for example, up to 3 million migratory birds are trapped or shot each year.[12] In Central and South America, turkey-like birds called guans and curassows are among the first animals to disappear when hunters invade remaining pristine forest areas.[13] The same holds true for 22 localized Asian pheasant species.[14] Parrots, while not widely hunted, are being loved to death instead: almost a third of the world's 330 parrot species are threatened by habitat loss and capture for the cagebird trade and habitat loss.[15]

Another form of wildlife exploitation—longline fishing—inadvertently kills hundreds of thousands of seabirds, which are hooked on baited lines and then drowned. More than 30 countries have longline fleets.[16] Although mitigation measures can dramatically reduce bird kills, little action has been taken. In 2001, seven countries signed an agreement under the Convention on the Conservation of Migratory Species of Wild Animals (known as the Bonn Convention) that will legally bind them to reduce longlining bycatch.[17] The agreement is awaiting ratification.

Pollution threatens birds in the oceans, near industrial sites, and in the countryside. Large and small oil spills kill many birds, including penguins, murres, and gannets. In addition to increased tanker traffic, aging vessels and lax restrictions make the business of transporting oil even more hazardous.[18] On land, oil and natural gas exploration, extraction, and pipelines threaten some of the world's most bird-rich habitats in Peru, Ecuador, and elsewhere.[19]

PCBs and other industrial effluents likely disrupt birds' endocrine systems and compromise their ability to attract mates.[20]

Agricultural pesticides kill millions of birds, weaken others, and deplete birds' food supplies of insects and wild plants.[21]

Another chemical threat, lead poisoning, weakens and then kills birds that swallow hunters' spent shot. Recent bans on lead shot in wetlands likely save many waterfowl from poisoning in the United States and a number of other countries.[22]

Skyscrapers, communications towers, and power lines can kill millions of migrating birds, which collide with the structures.[23] These threats, along with growing wind farm networks, require careful study to determine safe locations and heights that will minimize bird collisions.[24] To date, little work has been done in this area.

To these threats due to human activities must now be added another one: climate change. Earlier bird breeding, migration arrivals, and some bird species' northward range expansions seem to indicate the early effects of this. Also, some long-distance migrants now return to nesting grounds too late to exploit sped-up peaks in insect food supplies.[25] Climate change will likely alter many bird habitats in coming decades, perhaps causing the demise of isolated species.[26]

Decades of dedicated study lie ahead before scientists fully understand bird ecology, distribution, and behavior. Unfortunately, in many parts of the world, birds do not have much time. They will be lost if society doesn't protect their habitats and address other threats. Approximately a quarter of all bird species have ranges that are at most the size of Costa Rica.[27] More than half of these species are threatened or near-threatened, and for many, only fragmented habitats remain.[28]

In recent years, conservationists have catalogued 7,000 important bird areas (IBAs)—critical bird breeding and migration stopovers in 140 countries.[29] Also, hotspots for restricted-range and endemic species have been pinpointed in 218 endemic bird areas (EBAs), most of them in the tropics. Some IBAs and EBAs already include designated reserves. But many others remain unprotected and poorly studied.[30]

While reserve areas remain vital to protecting bird habitats, much of the world's land sits in private hands. Community, corporate, and government involvement in varied conservation efforts will be required to ratchet biodiversity and bird conservation to a higher status as part of a sustainable strategy for the planet. Fortunately, in a growing number of cases, enterprise and environmentalism prove mutually beneficial. Habitat-friendly agricultural programs in the Netherlands, the United Kingdom, and the United States bear this out, as do a growing number of corporate efforts to conserve birds.[31]

The ecotourism industry can also provide benefits to local communities near critically important habitats. An increasing number of tour operators hire local guides to assist their tours.[32] Meanwhile, town-based guide-training programs are taking flight in South Africa and elsewhere.[33]

Although some bird extinctions seem imminent, many can be avoided with a deep commitment to bird conservation as part of a sustainable development strategy. As the world works toward a more balanced future, keeping an eye on birds will help us keep ourselves in check—if we care to heed the warnings.

Table 1: Conservation Status of the World's Birds, 2000

Status	Species
	(number)
Extinct in the wild[1]	3
Critically endangered	182
Endangered	321
Vulnerable	680
Near-threatened	727
No threatened status	7,884
Total	9,797

[1] Species no longer survives in the wild, but at least some individuals remain in captivity.

Source: Stattersfield and Capper.

Small Islands Threatened by Sea Level Rise

David Taylor

Sea level, like the weather, varies considerably from year to year for island nations.[1] (See Table 1.) A combination of many factors, including wind, ocean currents, ocean temperature, and periodic oceanic oscillations like El Niño, bring about this annual variation.[2]

Long-term trends make it clear that for most islands, as for the world in general, the sea is rising. In the twentieth century, global sea level rose 10–20 centimeters, averaging 1–2 millimeters per year.[3] The sea level rises from melting continental ice masses and from the expansion of the oceans due to climate change.[4]

LINKS pp. 40, 50, 92, 102

Over the next century, global sea level rise is expected to accelerate. According to the Intergovernmental Panel on Climate Change (IPCC), the sea level will rise 9–88 centimeters in the next 100 years, with a mid-estimate rise of 50 centimeters.[5] This translates into 5 millimeters per year—two to four times faster than during the twentieth century.[6]

In terms of culpability for global sea level rise, the small-island states are beyond reproach; in terms of vulnerability, they are the most at risk. Although plagued by their own internal environmental problems, these small nations account for less than 1 percent of global greenhouse gas emissions.[7]

Accelerated sea level rise brings up the possibility that, for the first time in history, an entire sovereign country could be lost due to environmental change.[8] The height of low-lying atolls, like those in the Pacific and Indian Oceans, rarely exceeds 2 meters, with maximum heights of 3–4 meters.[9] New Zealand has drawn up a plan to accept immigrants from the tiny Pacific island country of Tuvalu, where residents fear losing their homes to future sea level rise.[10] And the Indian Ocean nation of the Maldives—65 percent of which is less than 1 meter above sea level—has evacuated residents from four of the lowest lying islands to larger ones over the past few years.[11]

One study notes that the impact of sea level rise on the Marshall Islands, Tuvalu, and Kiribati would be "profound," including disappearance in the worst scenario; the impact on the Federated States of Micronesia, Nauru, and Tonga would be "severe," resulting in major population displacement; and the impact on Fiji and the Solomon Islands would be "moderate to severe."[12] Indeed, in 1999, Kiribati lost two uninhabited islets, Tebua Tarawa and Abanuea, to the sea.[13]

Sea-level-rise scenarios have been compiled for a few small-island states, with most focusing on the impact of a 1-meter rise—the "worst-case scenario" for the next 100 years. Such a rise would inundate or erode 940 hectares in Antigua, 1,000 hectares in Mauritius, 3,700 hectares in Tonga, and 340 hectares in Nevis.[14] A recent study calculated that a 1-meter rise in the Caribbean would inundate 98 coastal communities in Cuba, threatening more than 50,000 persons.[15] Majuro Atoll, in the Marshall Islands, would lose 8.6 percent of its total land area with a rise of this magnitude, and 12.5 percent of Betio Island, Kiribati, would be vulnerable to annual flooding.[16]

While the long-term threat to these islands is inundation, the more immediate and pressing problems are those associated with storm surges, flooding, coastal erosion, saltwater intrusion into freshwater supplies, coral bleaching, and economic attrition.

Storm patterns are heavily connected to local weather patterns, most notably El Niño, strongly affecting islands in the Caribbean and the Pacific. According to the IPCC, the warm episodes of El Niño, which affect rainfall and periods of drought for small-island states, have been more "frequent, persistent and intense" since the mid-1970s.[17] Cyclones cause storm surges, which can reach up to 6 meters in height. With elevated sea levels, these surges are predicted to be more destructive, and even more so if cyclone intensity increases due to climate change.[18]

Individual studies suggest an increase of 10–20 percent in the intensity of tropical cyclones under enhanced atmospheric carbon dioxide conditions.[19] In June 1997, Cyclone Keli destroyed an islet of Tuvalu—Tepuka Savilivili—washing away all the vegetation and rendering the islet uninhabitable.[20] Some flood-risk models suggest that by 2080, the number of people facing severe floods in the Caribbean, Indian, and Pacific Ocean regions would be 200 times higher than if there were no sea level rise.[21]

Accelerated coastal erosion has caused some beaches in Trinidad, where sea level has risen four to eight times faster than the global average, to retreat by as much as 2 meters a year during the past 15 years.[22] In Fiji, where sea level has risen at the average global rate, beaches at Viti Levu and Taveuni have retreated by about 75 centimeters per year.[23] Coral reefs, which provide sand and a buffer for the beach, suffer severe impairment or death with ocean temperatures of about 1 degree Celsius higher than the summer maximum. This condition, called bleaching, will be highest in the Caribbean and lowest in the central Pacific in the next few decades.[24]

Tourism is one of the most important economic sectors for island states. For a number of these countries, such as Antigua and Barbuda, the Bahamas, Barbados, Cyprus, Grenada, Jamaica, the Maldives, Malta, St. Kitts and Nevis, Samoa, and the Seychelles, tourist revenue makes up more than 20 percent of the gross national product.[25] In addition to the degradation of natural resources, equatorial islands worry that global warming will lead to milder winters in industrial countries in northern latitudes, decreasing the incentive to travel for a large number of tourists.[26]

Another economic concern for islands is the reduction of their exclusive economic zones (EEZ), which provide sovereign development rights over 370 kilometers (200 nautical miles) of ocean area surrounding the islands.[27] These nations typically include tens to thousands of islets; for some mid-Pacific states, the EEZs are a thousand times larger than the land areas.[28] Even if they are uninhabited, disappearing fringe atolls could lead to a reduction of the EEZ and, therefore, a reduction in fishing license revenues for the government.

Table 1: Current and Historical Sea Level Rise in Selected Island Countries

Country	Average Sea Level Rise, 2002	Long-Term Sea Level Rise
	(millimeters)	(millimeters per year)
Cook Islands	12	2.3
Fiji	2	4.0
French Polynesia	24	2.5
Galapagos	52	1.5
Japan	6	3.2
Kiribati	35	−0.2
Maldives	8	–
Saipan	6	–
Seychelles	6	–
Tonga	40	4.9
Tuvalu	38	0.9

Source: University of Hawaii, Permanent Service for Mean Sea Level, and the South Pacific Sea Level and Climate Monitoring Project.

Economy Features

Courtesy the SeaWiFS Project, NASA/Goddard Space Flight Center, and ORBIMAGE

Rich-Poor Divide Growing
Gap in CEO-Worker Pay Widens
Severe Weather Events on the Rise

Rich-Poor Divide Growing

Radhika Sarin

In 1960, the per capita gross domestic product (GDP) in the 20 richest countries was 18 times that in the 20 poorest countries, according to the World Bank.[1] By 1995 the gap between the richest and poorest nations had more than doubled—to 37 times.[2]

To a large extent, these vast income gaps drive global consumption patterns. Disproportionate consumption by the world's rich often creates pollution, waste, and environmental damage that harm the world's poor. For example, growing demand for fish for non-food uses, mainly animal feed and oils, is diminishing the source of low-cost, high-protein nutrition for a billion of the world's poor people.[3] Carbon dioxide emissions, about 60 percent of which come from industrial countries, are threatening the very existence of poor island nations and the agricultural productivity of many developing ones.[4]

LINKS pp. 44, 90, 108

Between 1980 and the late 1990s, inequality also increased within 48 of 73 countries for which good data are available, including China, Russia, and the United States.[5] These 48 nations are home to 59 percent of the world's population and account for 78 percent of the gross world product.[6] This trend contrasts sharply with earlier declines in the gap between rich and poor in a number of countries between the 1950s and the early to mid-1970s, a period of stable global economic growth.[7]

Inequality remained constant in 16 countries and decreased in only 9: France, Norway, the Bahamas, Honduras, Jamaica, Malaysia, Tunisia, South Korea, and the Philippines.[8] Recent data, however, suggest that inequality may have increased since 1998 in the latter two nations in the wake of the East Asian financial crisis, as well as in four nations where it had been constant: Brazil, India, Indonesia, and Tanzania.[9]

The most dramatic surges in inequality have occurred in nations in transition from Communist rule to market-based economies.[10] Like other countries in the region, Russia is struggling with rising poverty, unemployment, and violence.[11] The main driver of inequality has been "state capture"—the manipulation of government by firms and powerful individuals to create laws and regulations to their own advantage.[12] This has concentrated power in the hands of the elite, while the vast majority of Russians remain politically and economically disenfranchised.[13]

Many industrial nations, including New Zealand, Japan, and the United Kingdom, have also experienced increases in inequality since the 1980s.[14] This is correlated with declines in the minimum wage, lower unionization, the decreasing power of unions, and a widening gap in the wages of skilled and unskilled workers.[15]

Of all high-income nations, the United States has the most unequal distribution of income, with over 30 percent of income in the hands of the richest 10 percent and only 1.8 percent going to the poorest 10 percent.[16] Data from the U.S. Census Bureau indicate increases in household income inequality between 1968 and 2001, which follow decreases between 1947 and 1968.[17] In particular, the richest 5 percent of the population has experienced the greatest percentage gain in income, and within that group, the top 1 percent gained more than the next 4 percent.[18]

Inequality is not restricted to personal incomes. Health and education—two important indicators of well-being—reveal stark disparities among the world's "haves" and "have-nots." Despite numerous international commitments to closing the gaps in access to education and health care, these remain correlated with income levels.[19] For example, the infant mortality rate in low-income countries is 2.5 times greater than in middle-income countries and 13 times greater than in high-income countries.[20]

And national averages only illustrate one level of disparity. A study of 44 developing nations found that infant mortality in the poorest fifth of the population is on average about twice the level in the richest fifth.[21] Even in a relatively wealthy developing nation such as Turkey, infant mortality among the poorest fifth is about four times higher.[22] In the United States, significant differences in infant mortality between racial and ethnic groups are largely the result of disparities in socioeconomic status and access to health care.[23] The infant mortality rate among American Indians and Alaskan natives is 1.5 times the figure among whites, while that of African-Americans is 2.5 times higher.[24]

Income, health, and education can, in fact, reinforce one another, with higher income leading to better health and education and those, in turn, leading to higher income.[25] These relationships are strongest at the lower end of the economic spectrum. That is, small increases in the income of the poor can yield dramatic health and education benefits.[26]

The links between inequality, economic growth, and poverty are complex. Economic growth plays an important role in reducing poverty, but existing inequalities can hamper this. The share of income earned by the poor in an unequal society is low, so only a small share of the income generated by growth will benefit this group. Evidence confirms that growth reduces poverty by nearly twice as much when inequality is low than when it is high.[27]

The nature of growth, and particularly the way that additional income generated by growth is distributed, is another important determinant of the impact on poverty. At any given rate of growth, poverty will fall faster in countries where growth raises the incomes of the poor by more than it increases average income. This essentially means that poverty falls faster when growth is accompanied by decreases in inequality and slower when accompanied by increases in inequality.[28]

In Bangladesh, for example, per capita GDP grew about 2 percent a year during the 1990s, but the decline of poverty has been slower, and rural poverty in particular remains very high.[29] Had it not been for rising inequality between 1992 and 1996, the poverty rate in 1995–96 would have been about 7–10 percent lower than it actually was.[30]

Poverty alleviation also depends on equality in access to opportunities. If poor people have no access to income-generating opportunities because of a lack of education, training, mobility, or credit, growth is unlikely to benefit them. This also holds true for other segments of society that are discriminated against in access to resources: women, ethnic minorities, and indigenous groups.[31] Faster growth in the urban sector than in the rural sector—the larger source of poor people's income—can also exacerbate inequality, as happened in China between the mid-1980s and mid-1990s.[32]

High inequality can itself worsen poverty by lowering overall growth. An unequal society is prone to political instability, increased crime, and dysfunctional or easily toppled institutions.[33] Unequal access to education, credit, and other resources is also inefficient for society as a whole because it blocks marginalized groups from increasing their productivity.[34]

A recent analysis by economists at the United Nations University concluded that international poverty-reduction targets cannot be achieved at current levels of inequality, despite projected economic growth.[35] The Gini index of income inequality measures the extent to which the distribution of income (or of consumption expenditures) deviates from a perfectly equal distribution. A value of zero indicates perfect equality, while 100 represents perfect inequality. When the Gini index is higher than 40, growth and poverty reduction tend to be dampened.[36] In 55 countries around the world, the Gini index is above this threshold.[37] (See Table 1.)

Table 1: Income Inequality in Selected Countries, 1990s

	Share of Income		
Country	Poorest 20 Percent	Richest 20 Percent	Gini Index
	(percent)		
Denmark	9.6	34.5	24.7
India	8.1	46.1	37.8
United States	5.2	46.4	40.8
Russia	4.4	53.7	48.7
Zambia	3.3	56.6	52.6
Brazil	2.2	64.1	60.7

Source: World Bank. Data are for most recent year available.

Gap in CEO-Worker Pay Widens

Gary Gardner

The difference between the compensation of corporate chief executive officers (CEOs) and the pay of factory workers is gaping and growing steadily in the United States. In 2001, executives of surveyed corporations in the United States made more than $11 million—some 350 times as much as the average factory worker.[1] (See Table 1.) And this earnings differential grew more than fivefold between 1990 and 2001.[2]

Today, the U.S. gap is at least 10 times greater than the differential in other industrial nations, where tax laws and cultural norms have prevented huge increases in executive pay. But as U.S. executive compensation practices—which rely heavily on stock, rather than cash, as the primary form of CEO compensation—are adopted elsewhere, the earnings gap in other countries could increase as well.

LINKS pp. 44, 88

The average executive compensation of $11 million in the United States compares with the average pay of factory workers of $31,260.[3] Because earnings of manufacturing workers increased only 42 percent between 1990 and 2001, the bulk of the fivefold increase in the pay gap clearly came from sharp increases in executive pay. The pay for manufacturing workers in other industrial countries grew on average 39 percent, roughly similar to that of workers in the United States.[4]

Growing pay discrepancies in the United States emerge largely from a compensation system skewed in favor of the CEO. In the past two decades, the shift to compensation based on stock options gave CEOs the right to buy company stock in the future at a price set today. The assumption was that tying executive compensation to company stock would give CEOs a strong incentive to ensure that a company is financially successful.

In the rapidly rising stock market of the 1990s, executives often held on to their options, allowing the value of the stocks to rise, then purchased the stock at the low price they had locked in. This led to huge accumulations in wealth. In the year 2001 alone, the average value of stock options received by CEOs in 370 surveyed U.S. companies was at least $8 million, more than triple the average level of CEO salary and bonuses ($2.3 million).[5] In the 1990s, the generous grants of options quickly skyrocketed in value. In an extreme example, Oracle CEO Larry Ellison cashed in $706 million in long-held options in 2001, making his take-home pay that year more than 17,000 times greater than the pay of the average manufacturing worker.[6]

With the fall of the stock market in 2001, executives saw an average decline of 43 percent in pay-related wealth (the sum of pay, bonuses, and accumulated stock options).[7] But two things cushioned this decline and give lie to the claim that compensation is closely pegged to company performance. First, pay packages are typically determined by committees of other CEOs, making the setting of compensation levels a political rather than an economic process—and a process that CEOs often influence.[8] Second, when stock prices fall, many compensation committees simply give the CEO more stock options, or they trade deflated stocks for newer and more valuable options.[9] Thus Ellison's $706-million record bonanza in 2001, for example, came even as his other stock holdings lost more than $2 billion in value.[10]

The use of stock options outside the United States, though less common, is on the rise as well. In the mid-1990s, stock option plans were allowed in only 10 of 26 nations surveyed.[11] By 2000, 19 of them had such schemes.[12] In Japan, where options have been allowed only since 1997, the number of firms using them grew by 7 percent between 1999 and 2000.[13] The trend appears to be one of the consequences of a globalization of the executive labor force, as corporations compete across borders for managerial talent. This can lead to odd situations in which branch executives working in the United States make substantially more than their bosses at headquarters in Europe or Asia.

In the United States, options are attractive not only to CEOs but to corporations as well, because they open an accounting loophole that saves companies money. U.S. accounting rules do not treat options as expenses, so millions of dollars in executive compensation do not show up on corporate balance sheets. Indeed, the difference between the income that corporations report to shareholders and what they report to the Internal Revenue Service grew by 70 percent just in the late 1990s, as more corporations took advantage of this loophole.[14] It is as though companies found a way to pay their executives at no cost to the corporation.[15] This accounting trick artificially inflates stock values—by 5–8 percent, in U.S. companies, according to a study by the U.S. Federal Reserve.[16]

Indeed, because options are profitable to CEOs and corporations alike, they now account for about 80 percent of executive compensation packages in the United States.[17] Elsewhere the figure is lower, but growing: in Australia, options and other long-term incentives made up 35 percent of CEO compensation in 1998, up from 13 percent in 1990.[18] Options are so liberally dispensed in the United States that they now account for about 15 percent of all stocks outstanding in the country.[19]

Such a system of compensation puts CEOs at odds with workers, both directly and indirectly. A 2002 study by the Institute for Policy Studies shows that many of the companies under investigation for accounting irregularities saw their CEOs claim lavish salaries—in part because of options—even as workers were laid off by the thousands.[20] U.S.-style compensation packages put jobs in jeopardy because they encourage executives to take excessive risks that will inflate stock values and because aggressive accounting methods overstate company earnings, a charade that cannot continue indefinitely.[21]

As taxpayers, workers are also hit.[22] While companies do not report stock options on the balance sheets in their annual reports, they do report options in their tax returns, as this reduces their tax liability—and increases the tax burden of the rest of society. And to the degree that workers hold stock, they are also hit, because liberal grants of stock options dilute the value of stock for other shareholders.[23]

Table 1: CEO-Worker Disparity in Pay, Selected Countries, 2001

Country	Average Annual Pay: Chief Executive Officers	Average Annual Pay: Manufacturing Workers	Ratio
	(dollars)		
United States			
With stock options	10,926,000	31,260	350
Without options	1,932,580	31,260	41
Mexico	866,833	3,720	233
Brazil	530,220	4,840	110
Italy	600,319	19,880	30
Canada	787,060	27,040	29
Spain	429,725	16,140	27
France	519,060	21,500	24
United Kingdom	668,526	27,720	24
Australia	546,914	23,460	23
Netherlands	604,854	29,000	21
New Zealand	287,345	15,300	19
South Korea	214,836	11,940	18
Sweden	413,860	28,960	14
Japan	508,106	36,960	14
Germany	454,974	34,760	13
Switzerland	404,580	35,180	12

Source: Towers Perrin, *Business Week*, and Bureau of Labor Statistics.

In response to recent options-related American corporate scandals, reform of options packages are being considered, even on Wall Street.[24] Many major corporations are now expensing their grants of stock options, and others are looking to tie stock grants more directly to company performance. At the same time, there are efforts to limit the gap between the pay of workers and that of executives, or at least to end government support for huge gaps. A bill before the U.S. Congress, the Income Equity Act, would change the U.S. tax code to cap the deduction for executive pay at 25 times that of the lowest paid worker in the company.[25]

Severe Weather Events on the Rise

Janet L. Sawin

The year 2002 set numerous local and regional records for windstorms, rain intensities, floods, droughts, and temperatures. Economic losses from weather disasters worldwide approached $53 billion, a 93-percent increase over 2001 losses.[1] (See Figure 1.) The increase was due in part to the return of El Niño in mid-2002.[2] The number of natural disasters totaled about 700; of these, 593 were weather-related events.[3] Windstorms and floods accounted for 98 percent of total 2002 insured losses from natural catastrophes.[4]

Weather disasters also took a significant human toll. Nearly 8,000 people died in storms, floods, droughts, heat waves, or extreme cold.[5] Many who survived faced the threat of diseases, including cholera, dysentery, malaria, and dengue fever.

LINKS pp. 28, 40, 66, 102

The highest costs of weather disasters, in dollar terms, are borne by industrial nations.[6] But developing countries suffer far higher losses as a share of their gross domestic products, as well as the majority of fatalities.[7]

The most costly event of 2002 was the flooding of the Danube and Elbe Rivers in August. Munich Re, a reinsurance company that compiles data on global disasters, called them "the worst floods in Europe for centuries, probably since the millennium flood in August 1342."[8] In less than two days, Germany received as much rain as it normally gets in a year.[9] At least 108 people died and 450,000 were forced to evacuate.[10] Total economic losses were estimated at $18.5 billion.[11]

Extreme cold in Moscow killed more than 300 people in December and early January, while eastern Russia had its worst snowstorm in at least 50 years.[12] Melting snows led to record floods, forcing thousands from their homes in southern Russia.[13] The following month, Bolivia's capital, La Paz, was hit by the most powerful storm in its history—receiving almost a gallon of water per square foot in less than an hour.[14] Then heavy snowfall and extreme cold hit Bolivia, Peru, and Argentina in July, killing at least 59 people and affecting 86,682 in Peru alone.[15]

In May and June, Southwest Asia sweltered in temperatures as high as 50 degrees Celsius (122 degrees Fahrenheit).[16] More than 1,200 people died in India, the highest one-week death toll on record for heat waves there.[17] In late July, torrential rains, mudslides, and floods killed nearly 300 Indians and affected more than 10 million.[18] Yet at the same time, expected monsoon rains neglected much of the country, causing the first all-India drought in 15 years.[19]

Warm, dry weather contributed to massive wildfires and agricultural losses in the United States and Australia.[20] Many U.S. regions experienced the worst drought since the Dust Bowl.[21] Australia lost some 40,000 rural jobs between July and October due to drought, and analysts expect economic growth to drop 1 percent as a result.[22] China also suffered major losses, estimated at $1.2 billion, due to the most severe drought in over a century.[23] More than 800,000 people in eastern and northern China were affected.[24]

Heavy rains in Kenya killed at least 53 people and displaced more than 150,000 in May.[25] As Kenya battled floods, the Eritrean government reported a drought that was unusually bad even for that country.[26] While AIDS, war, and other political problems also play a role, erratic weather patterns are the prime cause of famine for upwards of 18 million people across Africa.[27]

After months of dry weather, typhoons hit much of Southeast Asia and Japan in mid-2002.[28] Record-breaking rains, floods, and landslides killed nearly 1,100 people and injured more than 80,000 in China from June through August.[29] The flooding of Hunan's Dongting Lake affected 8.4 million people and cost more than $5.4 billion.[30] In September, Typhoon Rusa struck South Korea, setting a national record for rainfall and damage and costing the nation $6.6 billion.[31] Halfway around the globe, September was also the most active tropical storm month on record for the Atlantic basin.[32]

Since 1980, a total of 10,867 weather-related disasters have caused more than 575,000 deaths and entailed costs of more than $1 trillion (in 2001 dollars).[33] The frequency of severe weather events is clearly on the rise. In the United States, the number of weather-related disasters has increased fivefold since the 1970s.[34] Worldwide, the number of big weather catastrophes has quadrupled since the 1960s.[35]

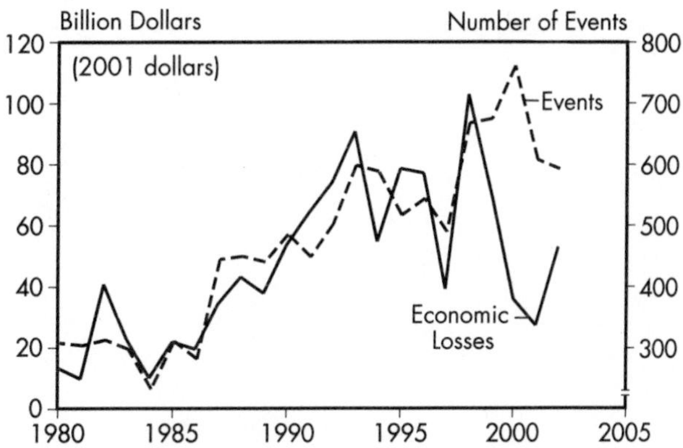

Figure 1: Number and Cost of Weather-Related Disasters, 1980–2002

As a result, average annual losses from weather events are rising as well, from more than $25 billion in the 1980s to nearly $71 billion in the 1990s.[36] Losses due to great catastrophes—which overtax a region's resources, making external assistance necessary—have increased even faster. Average annual losses from catastrophic weather events exceeded $43 billion during the 1990s—triple the figure in the 1980s, five times that of the 1970s, and eight times the average in the 1960s.[37]

Although the average number of deaths per weather event falls with improvements in forecasting and preparedness, the total number of people affected is rising.[38] In the Oceania region, for example, the number of deaths due to weather disasters rose by 21 percent between the 1970s and 1990s, while the number of people affected swelled from 275,000 in the 1970s to 18 million in the 1990s.[39] Environmental disasters—including severe weather—are to blame for 58 percent of the world's 43 million refugees.[40] Klaus Töpfer of the U.N. Environment Programme (UNEP) believes that the number of environmental refugees could double to 50 million by 2010.[41]

These economic and human costs have multiplied over the years due to not only the surge in extreme weather events but also rising global population and increasing concentrations of people and wealth in urban areas and vulnerable regions. Human activities such as clearcutting of upstream slopes have increased the impacts as well. In many cases, efforts to avert or lessen disaster, such as construction of dikes, dams, and avalanche barriers, have drawn people to coastal areas, riversides, and hillside locations, giving them a false sense of safety and perversely increasing the costs of future weather-related disasters.[42]

Scientists believe that rising global temperatures may increase the intensity and frequency of extreme weather events even more.[43] During the twenty-first century, average global surface temperatures are projected to increase at a rate unprecedented over at least the past 10,000 years.[44] Even slight temperature increases can shift low-pressure systems from their usual paths, causing sudden and significant increases in the frequency of heavy rainfall in a particular area.[45] Small increases in event severity can lead to multiple increases in damage and costs—for example, a 10-percent increase in wind speed can increase damage by 150 percent.[46] Thus climate change is expected to exacerbate the upward trends of economic and human costs.

A recent UNEP report concluded that if current trends continue, economic losses from natural disasters will reach $150 billion annually within the next decade.[47] According to experts at Munich Re, some single "worst case" disasters could exceed $100 billion.[48] Rising costs could stress insurers and banks to the point of insolvency.[49]

Resource Economics Features

Jeff Vanuga, USDA Natural Resources Conservation Service

High Farm Subsidies Persist

Harvesting of Illegal Drugs Remains High

High Farm Subsidies Persist

Brian Halweil

The governments that belong to the Organisation for Economic Co-operation and Development (OECD) gave $311 billion in subsidies to their agricultural sectors in 2001 (the last year for which data are available), which was down from $329 billion (in 2001 dollars) in 2000.[1] (See Figure 1.) Three quarters of these subsidies went directly to farmers, while the remainder supported food welfare, agricultural research, and government agriculture departments.[2]

Governments subsidize farmers in two main ways: through direct payments (based on production, acreage, or head of cattle, for example) and through price supports for various commodities. Although politicians generally argue that these subsidies provide a social safety net for rural communities and assure domestic food security, the way in which subsidies are distributed can actually undermine rural areas.

LINKS pp. 28, 30, 98

Moreover, economists argue that these payments distort production and trade, since they encourage farmers to produce more than the market demands. At the same time, the least distorting payments—those targeted at poorer farmers or payments to encourage farmers to improve their environmental performance—remain just 1 and 3 percent of support in OECD nations, respectively.[3]

On average, farmers in OECD nations received 31 percent of their income from government subsidies in 2001, compared with 38 percent in the mid-1980s.[4] But the share varies widely—from 4 percent in Australia and just 1 percent in New Zealand, which eliminated almost all farm subsidies in the 1980s, to 60 percent or more in Iceland, Japan, South Korea, Norway, and Switzerland.[5] In the European Union (EU), farm payments account for 35 percent of farm income, compared with roughly 20 percent in Canada, Mexico, and the United States.[6]

The average OECD farmer received $12,000 in farm payments in 2001, ranging from a high of $35,000 in Norway to under $1,000 in Mexico, Poland, and Slovakia.[7] This measure is misleading, however, since in all countries the distribution is highly skewed toward the largest producers.[8] Because they are tied to production, the payments also tilt the table toward the largest and wealthiest farmers, putting smaller, family farmers at a competitive disadvantage.[9] A 1995 analysis showed that the largest 25 percent of farms in the European Union got nearly 90 percent of total support.[10] And the largest 10 percent of U.S. farms are due to receive two thirds of the estimated $125 billion in farm payments distributed over the next decade.[11]

Several nations have recently adjusted their agricultural policies to try to help small or disadvantaged farmers more.[12] For instance, Hungary increased per acre payments for small farmers, while keeping them the same for larger farms, and Turkey now gives payments on a maximum of 20 acres.[13]

Because most farm subsidies are tied to the production of a handful of commodities—such as corn, soybeans, and beef—the payments help encourage farms that are low in diversity and high in agrochemical use, and they inhibit the adoption of resource-conserving practices. Farmers interested in diversifying out of the few crops that receive support lose a significant source of income.[14] A study of cropping patterns in South Dakota over the past half-century found that federal subsidies that gave disproportionate support to corn, wheat, and soybeans encouraged less diverse fields and fewer rotations.[15]

Governments give other types of subsidies to agriculture—for irrigation water, fuel, and agrochemicals—beyond the payments included in the $311-billion OECD estimate. The U.S. government provides an estimated $5 billion in irrigation subsidies each year.[16] One analysis estimated that agriculture costs $250 billion more in annual subsidies worldwide in the form of soil erosion, pesticide pollution, and other "externalities" that farmers do not have to pay for.[17]

Instead of making payments to farmers directly, many nations support farmers indirectly by purchasing food to distribute to poor consumers at reduced prices through welfare programs, for instance, or school meals. The United States spent $34 billion on domestic food assistance in 2001.[18] In 2001, India spent

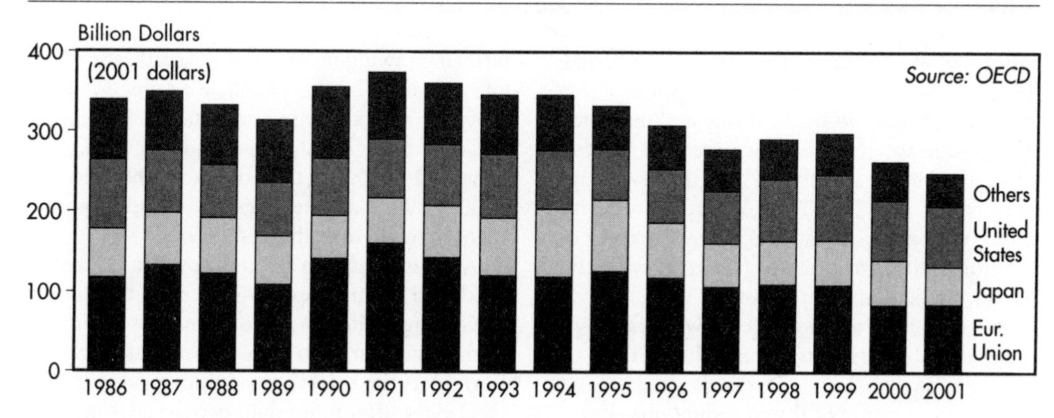

Figure 1: Total Agricultural Subsidies for OECD Nations, 1986–2001

$2.8 billion on subsidies used to buy food for government ration shops and public storage, in addition to $1.8 billion on fertilizer subsidies given to farmers and fertilizer makers.[19]

Even poorer nations have some subsidies, although none on the scale of the United States, Europe, and Japan. In 2001, Mexico paid farmers $6.5 billion in subsidies, and the government recently proposed a $10.25-billion subsidy package for 2003, partly to help farmers cope with subsidized U.S. crops.[20] In late 2002, in an effort to reduce herd sizes in the face of a regional drought, Namibia approved a program that would give farmers a $1.25 subsidy for each cow slaughtered—but this compares with $30–166 per cow in the European Union.[21]

Citizens finance farm subsidies not only through their taxes but also when price supports inflate food prices. In 2001, consumers in OECD nations paid roughly $137 billion more for food as a result of support for agriculture (lower than the $245 billion in the mid-1980s and $152 billion in 2000).[22]

Ironically, while subsidies can push food costs up domestically, nations and food exporters can dump the subsidized commodity on the world market and drive prices down, squashing local production in foreign nations.[23] Recent studies estimate that subsidies pull the price of U.S. and European crop exports to 20–50 percent below the cost of production, exerting substantial downward pressure on world market prices.[24] Beyond payments to farmers, the EU spent about $3 billion in 2001 on export subsidies (to reduce the price of exported goods and make them more competitive on the world market), and the United States spent roughly as much on export credit guarantees (to help foreign nations buy American farm goods).[25]

Despite the rhetoric about reducing subsidies as part of World Trade Organization negotiations on agriculture and in order to make farming "respond more to market signals," most OECD nations have made only minor adjustments to their farm subsidies.[26] In the United States, the 2002 Farm Bill will actually increase farm payments above previous levels.[27] In Europe, where farm subsidies make up half of the EU budget, governments appear to be gradually reducing farm subsidies per country in order to improve the environmental performance of agriculture and as part of the expansion of the Union.[28]

France is considering shifting 20 percent of all farm payments toward rural development and ecological farming programs in coming years.[29] France and Germany, the two biggest recipients of EU farm aid, will receive less aid over the next five years in order to finance payments to 10 new members starting in 2004.[30] And high-ranking Cabinet ministers in the U.K. government recently called for Common Agricultural Policy reform as a way to reduce the burden on taxpayers, improve the environmental performance of farms, and lower the damage to farm economies in developing countries.[31]

Harvesting of Illegal Drugs Remains High

Brian Halweil

Although production of the three major illegal drug crops—cannabis, coca, and opium poppies—is hard to track, the best global estimates indicate that it has increased dramatically since the 1980s, albeit with growth slowing in the last decade.[1] (See Table 1.) Products derived from these plants constitute over 95 percent of all global illegal drug sales.[2] (Sales of synthetic drugs, particularly amphetamine-type substances like "ecstasy," are growing rapidly but still only constitute a small share of all drug use.)[3]

The very nature of drug production—it is illegal, often takes place in unstable areas, and involves an industry that does not disclose its finances or activities—means that production estimates are generally based on indirect indicators such as drug seizures and drug treatment enrollment.[4]

LINKS pp. 74, 96, 120

Governments wrestle with the social paradox of drugs: they are a source of enjoyment and a lucrative business, but they cost billions of dollars each year in drug control and treatment of abusers. In a number of places—from Afghanistan to Colombia to Cambodia—drug crops have helped to finance violent conflicts, a reality that has prompted governments to militarize drug control and to criminalize drug use and production.[5] Yet eradication campaigns often take a heavy toll on the environment, while threatening the health and livelihoods of people in and near producing communities.

Cannabis is by far the most widely grown, sold, and consumed illicit drug. It is cultivated in an estimated 120 countries, compared with 35 countries where opium poppies are grown and just 6 with coca production.[6]

Coca—a bush whose leaves, which are used to make cocaine, have been consumed in raw form for thousands of years in the Andes—is grown primarily in Colombia, Peru, and Bolivia.[7] These three nations produce 98 percent of the world's cocaine, and Colombia alone is responsible for over 75 percent of global production.[8]

Coca production has declined slightly in recent years as a result of eradication efforts in Peru and Bolivia, partly offset by Colombian farmers growing more.[9] And the eradication effect tends to be short-lived, as production moves to adjacent regions or nations. For example, even though Colombian and U.S. authorities sprayed 260,000 hectares between 1994 and 2000, Colombia's total coca acreage grew more than threefold over the same period.[10]

Although Afghan farmers have traditionally produced opium poppies—the resin of which is used to make heroin, opium, and morphine—Afghanistan only surpassed Myanmar (formerly Burma) as the world's top opium producer in the 1990s, after more than two decades of war installed the crop as a major funding source for warring factions.[11] By the late 1990s, Afghanistan was responsible for over 70 percent of global opium production.[12]

In 2001, however, Afghan production plummeted by 94 percent—from 3,276 to 185 tons—after the ruling Taliban banned poppy cultivation.[13] As a result, production at the global level dropped by 65 percent, from 4,700 tons to 1,600 tons.[14] More recently, opium growers have taken advantage of the power vacuum created by the fall of Taliban regime and the U.S.-led war to once again make that nation the world's largest producer of opium poppies—with an estimated production of 3,400 tons in 2002.[15]

With recent instability in Afghanistan, the "Golden Triangle" of Southeast Asia, defined by Myanmar, Laos, and northern Thailand, has re-emerged as an important opium production center.[16] Myanmar and Laos grow 1,100 tons and 134 tons, respectively.[17]

Analysts estimate global illicit drug sales at between $300 billion and $500 billion each year, compared with just over $300 billion in annual drug sales for the pharmaceutical industry.[18] In some countries the illegal drug trade generates more money than any other single legal industry.[19] In Colombia and Mexico, for instance, drug exports rival revenues from oil, the top legal export.[20] Bolivia's estimated coca and cocaine exports in the early 1990s were half the size of the nation's total legal exports.[21] A 1998 estimate found that marijuana was the fourth most lucrative crop in the United States, after corn, soybeans, and hay, and the biggest

Table 1: Trends in Major Drug Crops, Circa 2000

Drug	Production Estimate	Major Producing Nations	Seizures Record
Cannabis	Some 120 countries cultivate cannabis. Annual production estimates vary widely—from 10,000 tons to 300,000 tons—with 30,000 tons considered a best guess.	Widely dispersed	4,500 tons of cannabis herb seized in 2000—twice as much as in 1990
Coca	Production of coca leaf has fluctuated at 300,000–350,000 tons, enough to yield 600 tons of cocaine in 1989, peaking at 950 tons in 1996, and reaching 827 tons in 2001.	Colombia, Peru, and Bolivia	335 tons of cocaine were seized in 2000, more than 90 percent in the Americas
Opium	Production climbed throughout the 1990s, from roughly 3,400 tons in 1989 to 5,800 in 1999. Opium production plummeted to 1,626 tons in 2001 but rose enough in 2002 to make about 160 kilos of heroin.	Afghanistan, Myanmar, Laos, and Thailand	75 tons of heroin and morphine were seized in 2000—five times as much as in 1990

Source: U.N. Office for Drug Control and Crime Prevention and U.N. International Drug Control Programme.

grossing crop in several states.[22]

Like other agricultural commodities, by far the largest profits in the drug business come at the retail end, with an estimated 90 percent or more of the final sale price going to local dealers and often a minuscule share going to the farmer.[23]

Despite this skewed distribution of profits, in many cases drug production, processing, smuggling, and retailing provide jobs and income in communities where there are few other opportunities.[24] In Mexico, many farmers are turning to opium or marijuana because their corn and other crops cannot compete with cheaper imported food.[25] In Baltimore, Maryland, drug dealing provides a stable—albeit dangerous—income in some neighborhoods where jobs are scarce.[26] In Afghanistan, revenue for each of the 200,000–250,000 families involved in poppy production is estimated at $3,000–4,000 a year, a substantial amount in one of the world's poorest nations.[27]

An estimated 2–5 percent of Peru's work force and between 8 and 17 percent of Bolivia's work force—in other words, hundreds of thousands of people—are directly employed in drug production or processing.[28] One analysis suggests this share approaches 50 percent in Colombia's centers of coca production.[29]

The drug business is so lucrative because of persistent high demand in wealthy nations. The United Nations Office for Drug Control and Crime Prevention estimates that 185 million people worldwide use drugs each year, roughly 4.3 percent of the population over the age of 15.[30] This includes at least 147 million marijuana users and roughly 13 million users of cocaine and heroin.[31] Use tends to be highest among men, single people, the unemployed, and people aged 15 to 35.[32]

North America and Western Europe remain the first and second largest markets, respectively, for illegal drugs.[33] Nonetheless, drug use is growing rapidly in Eastern Europe and the former Soviet states in addition to the developing world, where rising incomes and "spillage" from drug production and export have boosted local availability.[34]

Because most of the world's coca production and processing and most opium production and processing in Southeast Asia and Latin America occur in rainforests, some of the most acute costs of both drug production and drug control fall on the environment, when farmers clear forests to produce illicit drugs, when processors use toxic compounds, and when authorities spray herbicides during eradication campaigns.[35]

Health and Social Features

Jonathan Frerichs, Luthern World Relief

Number of Refugees Drops

Arunima Dhar

At the beginning of 2002, roughly one out of every 300 persons on Earth—19.8 million people in all—were classified by the U.N. High Commissioner for Refugees (UNHCR) as "people of concern."[1] Of this total, 12 million were officially recognized as refugees.[2] (The United Nations defines a refugee as a person who, "owing to well-found fear of being persecuted for reasons of race, religion, nationality, membership of a particular social group or political opinion, is outside the country of his nationality and owing to such fear, is unwilling to avail himself of the protection of that country.")[3] The other nearly 8 million included 940,800 asylum seekers, 462,700 returned refugees, 5.3 million internally displaced persons (IDPs), 241,000 returned IDPs, and 1 million others "of concern."[4] (See Table 1 for examples of displaced people.) The overall figure was 2 million below the previous year's total.[5] But another 50 million people were environmental refugees, IDPs not eligible for UNHCR aid, or otherwise uprooted people who were not counted by UNHCR.[6]

LINKS pp. 74, 84, 92, 120

Developing countries produced 86 percent of the world's refugees over the past decade, but at the same time they also provided asylum for 72 percent of the global refugee population.[7] Asia hosted the largest overall refugee population (5.8 million), while North America provided a home, even if temporarily, to an estimated 650,000 refugees.[8]

The number of people seeking asylum worldwide dropped slightly—from 1.1 million in 2000 to 923,000 in 2001.[9] Asylum applications in industrial countries, however, rose by 8 percent.[10] The United Kingdom received the largest number of applications in 2001 (92,000), followed by Germany (88,300) and the United States (83,200).[11] From 1992 to 2001, industrial countries resettled 1.2 million refugees, with the United States receiving 77 percent of all arrivals.[12] Afghanistan was the main country of origin of asylum seekers in 2001, with 66,800 Afghans applying for asylum in 144 countries.[13]

Some 463,000 refugees repatriated—returned to their original home country—in 2001, 40 percent fewer than in 2000.[14] The annual number of refugees who repatriated with direct assistance from UNHCR fell by 60 percent.[15] (But 2002 brought about a sharp reversal, as roughly 2 million Afghan refugees and internally displaced civilians returned to their homes in Afghanistan.)[16]

The UNHCR figures do not cover the entire population of uprooted people in need of aid. They do not, for instance, include an estimated 3.9 million Palestinians who are cared for by the U.N. Relief and Works Agency for Palestine Refugees in the Near East.[17]

Internally displaced people largely fall outside the core mandate of the UNHCR, primarily because they remain within their own borders. The U.S. Committee for Refugees puts the figure of internally displaced persons at 22 million, while the Norwegian Refugee Council's Global IDP Project estimates the number to be 25 million.[18]

Outnumbering international refugees two-to-one and receiving far less help and protection, IDPs pose a key humanitarian challenge.[19] Of the 20–25 million people thought to be internally displaced due to conflict or persecution, only 5.3 million receive UNHCR aid.[20] The latest estimates for mid-2002 placed the number of people displaced by armed conflict in Asia at over 4.6 million; Indonesia was the most affected, with 1.3 million internally displaced people.[21] Afghanistan and Sri Lanka each had about 1 million IDPs.[22] Nearly 2.2 million people were displaced in the Americas, twice as many as in 1996, due entirely to the escalating violence in Colombia, where 300,000 people are being displaced yearly.[23] In Europe, at least 3.3 million people have still not been able to return to their homes due to conflicts in 11 countries—outnumbering the official refugee population of 2.7 million.[24] But by far the largest IDP population is in Africa, with 10–13 million people displaced.[25]

The number of environmental refugees and those displaced by natural disasters and development projects has also risen considerably.[26] In 1995 (the latest year for which estimates are available), at least 25 million people fell into this

Table 1: Selected Examples of Displaced People Worldwide

Country or Region	Description of Displacement
Sudan	Approximately 4.4 million Sudanese were uprooted at the end of 2001—including an estimated 4 million internally displaced persons and some 440,000 Sudanese living outside the country as refugees and asylum seekers.
Afghanistan	Some 700,000 Afghans were newly internally displaced in 2001; 200,000 Afghans fled to Pakistan; and another 200,000 fled to Iran. At the same time, 140,000 Afghans were repatriated and 120,000 were deported.
Burundi	IDPs in Burundi account for a striking 10 percent of the population. Human rights groups have criticized the government for conscripting some 14,000 IDP children into the army.
Colombia	In 2001, some 342,000 Colombians were newly displaced by political violence and 13,000 applied for asylum in other countries.
Gaza Strip/West Bank	Some 26,000 Palestinians fled to Jordan from the Occupied Territories.
Bosnia	In Kosovo, 220,000 non-Albanians have had to move to other parts of the Federal Republic of Yugoslavia. Approximately 15,000 ethnic Albanians also continued to leave southern Serbia for Kosovo. Some 99,000 Bosnians returned to their places of origin—93 percent to areas where they are in the ethnic majority.

Source: U.S. Committee for Refugees, UNHCR, and U.N. Office for the Coordination of Humanitarian Affairs.

category.[27] And this number was projected to double by 2010, growing by almost 8,500 a day.[28]

The Intergovernmental Panel on Climate Change has predicted that a large share of Bangladesh's landmass could be submerged due to rising sea levels, turning millions of people into refugees.[29] Many of the residents of the small Pacific island nation of Tuvalu are expected to seek shelter in New Zealand in the coming decades, as they worry that rising seas will wash away their homes.[30]

In a sign of the disruption from natural disasters that could lie ahead, Hurricane Mitch displaced more than 1.2 million people throughout Central America, while the Venezuelan floods displaced about 150,000.[31] In Brazil, some 12 million people were affected by drought, forcing many of the rural poor to migrate to cities like São Paulo, whose population of 19 million in 1999 was swollen by 300,000 of the "wandering poor."[32]

According to the World Commission on Dams, some 40–80 million people were displaced from their homes by dams over the last half-century.[33] And as of 2000, about 300 development projects supported by the World Bank "adversely affected" 2.6 million people in 548,000 households through physical or economic displacement as a result of land acquisition for Bank-aided projects.[34]

The UNHCR also does not count people who live in "refugee-like situations," which means that typically they live in similar or worse conditions but do not meet the narrow definition of a refugee.[35] The U.S. Committee for Refugees estimated that approximately 4.4 million people were in this category at the end of 2002.[36]

Women account for 50–80 percent of the global refugee population.[37] Despite investigations pointing to a persistent problem of violence against and exploitation of refugee women and girls, an ongoing funding crisis has curtailed programs to prevent sexual and gender-based violence against female refugees. Funding by donor countries for international refugee programs was seriously inadequate during 2002. UNHCR expected to end the year some $170 million short of the $1.04 billion needed to address basic refugee needs.[38]

Alternative Medicine Gains Popularity

Danielle Nierenberg

Traditional medicine (TM) and what is called complementary/alternative medicine (CAM) are primary sources of health care for millions of people—about 80 percent of the world depends on traditional techniques alone for treating and curing illness.[1] In many parts of Africa, Asia, and Latin America, TM is the only health care many people receive; at the same time, the use of alternative medicine is growing in Australia, Europe, and North America.[2] (See Table 1.)

In Africa, about 80 percent of the population uses traditional remedies, while in China, TM accounts for 30–50 percent of health care.[3] In Canada, two out of three people have used alternative or traditional therapies at least once.[4] And in the United States, 42 percent of the population reports use of some traditional medicine, including medication (herbal, animal parts, or minerals) and other therapies (exercise or meditation).[5]

LINKS p. 68

According to the World Health Organization (WHO), TM refers to ways of protecting and restoring health that existed before the arrival of modern medicine.[6] It "incorporates plant, animal, or mineral-based medicines, spiritual therapies, manual techniques, and exercises... to maintain well-being, as well as to treat, diagnose, or prevent illness."[7] Complementary and alternative medicine can include TM, but it refers to a broad set of practices that are not based on a particular country's religious or spiritual traditions or that are not a part of its dominant health care system. Herbal remedies, spiritual therapy, acupuncture, yoga, various forms of indigenous medicine, and nutritional therapy can all be part of TM or CAM.[8]

Traditional and alternative medicine are increasingly used in part because of accessibility and affordability. People in poor nations can obtain them for free by gathering plants in forests and jungles or by growing them in gardens or between crops. In rural areas, traditional healers are also more readily accessible than doctors. In Tanzania, Uganda, and Zambia, the ratio of TM practitioners to the population is 1 to 200 or 1 to 400 compared with 1 to 20,000 for doctors trained in more modern medicine.[9] For many, TM is the only affordable source of health care for treating malaria, AIDS, and other diseases. According to WHO, most Africans living with HIV/AIDS use TM to both obtain relief and combat infections.[10] Treating malaria with herbal medicines is also far cheaper and can be highly successful.[11]

In industrial countries, many people decide to use TM or CAM because they believe these alternatives are less invasive or toxic than modern therapies.[12] In Japan, Australia, Canada, and Switzerland, 40–70 percent of people have used CAM, leading many conventional doctors to include some form of alternative medicine in their practices.[13] Acupuncture is the most popular CAM method used by conventional practitioners. In Belgium, for example, 74 percent of acupuncture treatment is administered by medical doctors.[14]

"Ecological medicine" is an emerging philosophy shared by many health care professionals, scientists, and environmentalists.[15] It adds an ecological component to non-toxic and natural medical practices. Epidemiologists have long acknowledged that cultural dimensions play a role in health. Now that human activities—population growth, resource abuse, a narrow pursuit of economic growth, and inappropriate technologies—are degrading the environment, there are new patterns of human and ecosystem poverty and disease.[16] And although intensive medical procedures, such as chemotherapy, have prolonged countless lives, many of these advances have come at great cost to both human health and the environment by creating medical waste and long-term contamination.[17] Ecological medicine advances public health by protecting the environment and creating a diverse, equitable, environmentally just system of guidelines for doctors and patients to follow when choosing treatments.[18]

Protecting the natural resources on which herbal and other traditional medicines are based is important, especially in developing countries that depend heavily on TM for basic health care.[19] Publicity about use of the African potato in treating AIDS, for instance, and growing demand for the plant are threatening its sustainable production in eastern and southern Africa.[20]

Unlike conventional therapies, the efficacy of many traditional and alternative treatments is not based on just scientific fact but is a holistic approach that emphasizes total health more than isolated symptoms.[21] Evaluation of TM and CAM "products" can be difficult. Plant identification, including the area of collection and the time of plant collection, is important in determining the effectiveness of therapies. And because a single plant can contain hundreds of natural constituents, it can be prohibitively expensive to establish which is responsible for an effect.[22] Similarly, cultural differences among practitioners and patients can create problems concerning the interpretation and application of a therapy.[23]

Governments are beginning to respond to the growing interest and use of TM and CAM by regulating them.[24] Some governments in Africa have started programs to improve knowledge among traditional birth attendants, as well as training doctors, nurses, and pharmacists about TM.[25] And in Mali, Nigeria, Rwanda, Senegal, and several other African nations, national budgets allocate funds specifically for TM.[26] Most developing nations also have national TM research institutes. In 2000, when the heads of 53 African countries signed the Abuja Declaration on Roll Back Malaria, they noted the importance of using TM to fight this devastating disease.[27]

In Norway, the government requires CAM practitioners to be certified.[28] The Norwegian Ministry of Health and Social Affairs also established a committee to study CAM's role in improving health care.[29] And in the United States, according to the newly established National Center for Complementary and Alternative Medicine, CAM funding has increased significantly—from just $2 million in 1990 to more than $68 million in 2000.[30]

Drug companies and institutions have also realized the economic opportunities that can be gained by tapping into the knowledge of traditional practitioners. Unfortunately, knowledge about traditional medicines is often adapted and patented by industry without the consent of its original custodians because they lack intellectual property rights.[31]

Although the use of traditional and alternative medicine is becoming more important as conventional health care costs rise, only 25 of WHO's 191 member states have developed a national TM/CAM policy.[32] In 2002, WHO established a traditional medicine strategy to evaluate methods, assure safety of products, train practitioners, and consider intellectual property issues.[33]

Table 1: Use of Traditional and Complementary/Alternative Medicine, Selected Countries and Regions

Country or Region	Description
China	Traditional Chinese medicine is fully integrated into China's health system and 95 percent of hospitals have units for traditional medicine.
India	Traditional medicine, such as *ayurveda, siddha, unani,* yoga, and homeopathy, is widely used in rural India, where 70 percent of the population lives.
United States	National expenditure on complementary or alternative medicine stands at $2.7 billion per year.
Japan	Seventy-two percent of registered western-style doctors use *kampo* medicine, the Japanese adaptation of traditional Chinese medicine.
Europe	Acupuncture is provided by 77 percent of the pain clinics in Germany; in the United Kingdom, 46 percent of doctors recommend patients get acupuncture elsewhere or perform it themselves.
Africa	Most Africans living with HIV/AIDS or malaria use some form of traditional medicine. In Ghana, drugs from a clinic can cost more than 10 times as much as self-treatment with herbs.
Central and South America	Regulation and registration of herbal medicines have been established in Bolivia, Chile, Colombia, Costa Rica, Ecuador, Guatemala, Honduras, Mexico, Peru, and Venezuela.

Source: World Health Organization.

Maternal Deaths Reflect Inequities

Radhika Sarin

An estimated 515,000 women die annually from pregnancy-related causes.[1] And for every maternal death, there are 30 cases of injury, illness, or disability, bringing the total number of women harmed each year by pregnancy or during childbirth to over 15 million.[2]

Of all health indicators, maternal mortality reveals the starkest disparity between industrial and developing nations.[3] Ninety-nine percent of maternal deaths occur in developing countries, where a large proportion of women give birth without the aid of skilled health personnel.[4] (See Figure 1.)

Four out of five maternal deaths are the direct result of complications arising from pregnancy and childbirth: hemorrhage, infection, unsafe abortion, prolonged or obstructed labor, and hypertensive disorders, such as convulsions.[5] The other 20 percent result from pre-existing conditions that are exacerbated by pregnancy, including anemia, malaria, hepatitis, and HIV/AIDS.[6] Poor nutrition, which often starts during childhood, and deficiencies of vitamin A, iodine, calcium, iron, and folic acid also contribute to poor maternal health.[7]

LINKS pp. 66, 68, 108, 110

The maternal mortality ratio—the number of maternal deaths per 100,000 live births—measures the risk of death once pregnancy has occurred. But since a woman faces this risk each time she is pregnant, a more comprehensive measure is the lifetime risk of maternal death, which also takes into account the average number of pregnancies per woman.

Women in countries with a high maternal mortality ratio as well as a high total fertility rate face the greatest lifetime risk. The chance of a woman dying during pregnancy or childbirth is as low as one in 5,500 in Australia and New Zealand and as high as one in 11 in the nations of Eastern Africa.[8] Compared with women in industrial countries, those in developing nations run a 40 times greater risk of maternal death during their lives, and those in the 49 "least developed" countries (using the standard U.N. designation) are more than 150 times as likely to die during their lifetime due to pregnancy or childbirth.[9]

Most maternal deaths could be avoided if not for the vast disparities in maternal health care both between and within countries. The presence of a skilled attendant with midwifery skills and the availability of transport to a health facility in case of an emergency are basic components of safe motherhood. Traditional birth attendants, whether trained or untrained, are often unable to handle serious complications.[10]

Skilled attendants deliver 99 percent of births in the industrial world, compared with 58 percent in the developing world and only 34 percent in the "least developed" countries.[11]

A World Bank study of 44 developing

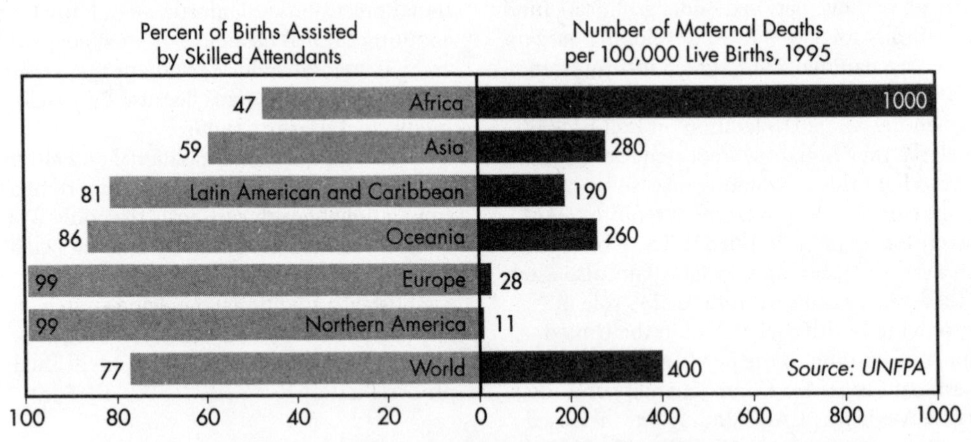

Figure 1: Skilled Care at Delivery and Maternal Deaths, Regional Comparisons

nations shows that personal income is strongly correlated to maternal health and care. In South Asia, for example, the richest fifth of the population is 10 times more likely to have skilled assistance than the poorest fifth.[12] The place of delivery is also linked to income: in the countries studied by the Bank, 80 percent of the poorest one fifth of women delivered at home while, in direct contrast, 80 percent of the richest one fifth delivered at health facilities.[13]

Simply meeting the demand for family planning services would lower maternal mortality by reducing unintended pregnancies. A recent analysis by the Global Health Council found that nearly 700,000 women died as a result of unwanted or unintended pregnancies between 1995 and 2000.[14] (The majority—more than 400,000—died from unsafe, unsanitary, and sometimes illegal abortions.)[15] This suggests that roughly 115,000 maternal deaths could be prevented every year through birth spacing or prevention.[16]

Regions with the most limited reproductive health care—sub-Saharan Africa and South Asia—suffer the greatest abortion-related maternal mortality, despite having lower abortion rates than other regions.[17] For instance, 13 percent of all pregnancies end in abortion in sub-Saharan Africa compared with 22 percent in Northern America.[18] Yet the rate of abortion-related maternal deaths per 100,000 pregnancies in sub-Saharan Africa is over 300 times greater than in Northern America.[19]

Trends in maternal mortality are difficult to evaluate. In recent years, new ways of measuring this important indicator of a nation's well-being have been developed, making it hard to compare recent data with data from past analyses. And even the best current estimates have large margins of error, adding to the difficulty of trends analysis. However, the percentage of births attended by skilled attendants serves as a good proxy indicator because it is highly correlated with maternal mortality and is easier to measure.

Trend data on skilled attendants at delivery are available for 51 nations, accounting for 74 percent of live births in the developing world.[20]

Table 1: Skilled Attendants at Birth, by Region, 1990 and 2000

Region	1990	2000
	(percent of births attended by skilled personnel)	
Sub-Saharan Africa	40	42
Middle East and North Africa	51	64
Asia	37	50
Latin America and the Caribbean	76	85
Total	42	53

Source: UNICEF.

(See Table 1.) Delivery care has increased in many regions over the past decade, but in sub-Saharan Africa, where maternal mortality is highest, there has been little improvement. In fact, in some African nations delivery care has worsened. In 1998–99, skilled personnel attended only 31 percent of births in Burkina Faso, down from 42 percent in 1993.[21] In Kenya, the proportion fell from 50 to 44 percent between 1989 and 1998.[22]

Improving maternal health care has a positive effect on infant and child survival as well. Eight million stillbirths and newborn deaths occur each year, largely as a result of insufficient care during pregnancy and delivery.[23] And research from Bangladesh shows that children under age 10 who have lost a mother are 10 times more likely to die than those whose mothers are still alive.[24]

Safe motherhood depends not only on providing sound medical care but also on correcting social and economic disparities between men and women. These disparities are often the root causes of women's poor health. Discrimination against women—in access to food, resources, and education—remains an obstacle to improving maternal health.

Consumption Patterns Contribute to Mortality

Erik Assadourian

In 2000, 55.7 million people died around the world, succumbing to a wide range of illnesses and conditions.[1] (See Table 1.) Cardiovascular diseases, including various chronic heart conditions and stroke, were the largest cause—killing 16.7 million people.[2] Infectious and parasitic diseases, including AIDS, tuberculosis and respiratory infections, malaria, and diarrheal diseases, were the second largest, taking 14.4 million people.[3] And cancers were the third, responsible for 6.9 million deaths.[4]

On the broadest scale, the two population groups at opposite ends of the income scale—the affluent and the impoverished—are dying from very different diseases. Infectious diseases primarily plague the developing world, especially people earning less than $2 a day, who cannot afford clean water, sanitation, or nutritious food.[5] People in Africa and Southeast Asia are the most gravely affected by these: they account for 75 percent of the deaths from infectious diseases, but just 36 percent of the world's population.[6] In contrast, cardiovascular diseases and cancers primarily affect those who consume too many unhealthy foods, tobacco, alcohol, and drugs, and who lead sedentary lifestyles—primarily Europeans and Americans.[7] They account for 42 percent of cardiovascular diseases and cancers, yet only 28 percent of the world's population.[8]

LINKS pp. 68, 70, 106

Underlying such overt causes as infectious and cardiovascular diseases are a number of risk factors for these illnesses. In 2002, the World Health Organization identified several major risks and assessed the contribution of each to global mortality.[9] Of course, risk factors do not act exclusively—for instance, diarrheal diseases can be caused jointly by poor sanitation and poor nutrition. Thus, adding risk factors results in high-end estimates. Even as such, the WHO analysis conveys the significant impact consumption has on mortality. Indeed, in 2000, overabundant consumption of resources accounted for up to 46 percent of mortality, while lack of access to resources accounted for up to 23 percent of deaths—roughly 99 percent of which occurred in the developing world.[10]

In the developing world, where people often lack access to clean fuels and well-ventilated shelter, the use of such solid fuels as coal, wood, and dung for cooking and heating caused 1.6 million deaths by triggering respiratory infections and lung diseases.[11] Unsafe sex, mainly through spreading HIV or a lack of contraception, killed 3 million people in 2000—75 percent of whom lived in Africa.[12] Due to lack of access to education, condoms, and health care, HIV is growing unchecked throughout this continent, with more than 3.5 million people newly infected in 2002.[13]

Dietary deficiencies, including lack of calories, protein, iron, zinc, and Vitamin A, produced up to 6.2 million deaths, mostly in children and women of reproductive age—primarily by weakening the immune system, thus increasing susceptibility to infectious diseases.[14] The lack of access to clean water and sanitation led to 1.7 million deaths in 2000, the vast majority from diarrheal diseases.[15] Of these deaths, 99.8 percent occurred in the developing world, and 90 percent of the victims were children.[16]

Improving the allocation of health resources and sanitation would dramatically reduce infectious disease deaths. Currently, almost 30 million of the 130 million children born each year do not receive vaccinations.[17] Immunizing every child would prevent 3 million deaths each year, while costing just $1.3 billion more than the world currently spends annually on vaccinations—far less than the costs of long-term treatment and disability.[18] Providing access to sanitation to just half of the 3 billion people who currently lack it would reduce the number of years of lost life by 30 million, at a cost of just $37.5 billion over 10 years.[19]

In the industrial world, deaths brought about by lack of access to resources accounted for just 1 percent of deaths.[20] People in these countries suffered primarily from diseases related to poor dietary and lifestyle behaviors.[21]

High blood pressure, high cholesterol, being overweight, and eating too few fruits and vegetables together caused up to 7.6 million deaths in industrial countries by increasing the risk for a number of diseases, including stroke, heart diseases, cancer, and diabetes.[22] These conditions are primarily triggered by a diet too high in salt, sugar, fat, and calories; as these increase in the diet—often in the form of processed foods—they displace healthier, less convenient foods, such as fresh fruits and vegetables.[23]

Physical inactivity exacerbates poor dietary behaviors and contributed to 855,000 deaths in industrial countries by increasing rates of heart disease, cancer, and diabetes.[24] The use of addictive substances only compounds the problems caused by poor diet and lack of exercise. Tobacco and alcohol use cause heart disease, stroke, and cancers and were responsible for 3 million deaths in the industrial world.[25]

These problems do not plague only the industrial world. More people die from overconsumption in developing countries (up to 14.3 million) than in industrial ones.[26] And even in high-mortality developing countries, where poor sanitation and dietary deficiencies account for up to 42 percent of deaths, overconsumption now accounts for up to 27 percent of mortality.[27] As conditions in the developing world improve, those living there often undergo a "risk transition": increases in income provide more access not only to food and clean water but also to processed foods and to tobacco, alcohol, and drugs; together these shift the disease burden from infectious to chronic diseases.[28] In low-mortality developing countries, where poor sanitation and undernourishment are less of a problem, overconsumption now causes up to 45 percent of deaths.[29]

A few countries have successfully countered the poor health that stems from the increases in unhealthful consumption that can accompany growing affluence. South Korea, for example, has minimized obesity by promoting its traditional diet—high in rice and vegetables and low in fats, salt, and sugar—through a combination of education, support for local farming, and mass media campaigns.[30]

Two decades after going through a nutritional transition in the 1950s, Finland suffered from one of the highest rates of cardiovascular disease in the world.[31] In the 1970s, the government worked with health experts, the food industry, and local communities to reverse this trend, and by 1995 the program had reduced heart disease deaths by 65 percent.[32]

Yet South Korea and Finland represent exceptional cases. Most governments have not faced the epidemic of overconsumption in their societies and will need to work aggressively if they are to prevent rapid growth in mortality in the coming decades.

Table 1: Global Mortality by Cause, 2000

Cause of Death	Number	Share of Total
	(thousand)	(percent)
Cardiovascular diseases	16,701	30.0
Infectious and parasitic diseases	14,398	25.9
Cancers	6,930	12.4
Maternal and perinatal conditions and congenital abnormalities	3,591	6.4
Chronic respiratory diseases	3,542	6.4
Unintentional injuries (such as auto accidents)	3,403	6.1
Digestive diseases	1,923	3.5
Neuropsychiatic disorders	948	1.7
Violence and war	830	1.5
Genitourinary diseases	825	1.5
Suicide	815	1.5
Diabetes	810	1.5
Nutritional deficiencies and disorders	669	1.2
Other	309	0.6
Total	55,694	100

Source: World Health Organization.

Orphans Increase Due to AIDS Deaths

Radhika Sarin

At the end of 2001, an estimated 13.4 million children under the age of 15 in Africa, Asia, and Latin America and the Caribbean had lost a parent to AIDS.[1] (See Figure 1.) More than 11 million of these "orphans due to AIDS" live in Africa.[2] (The United Nations defines "orphan" as any child under the age of 15 who has lost either one parent or both parents. Although the loss of one parent may be less of a burden on children and other relatives, it is still a cause of physical and emotional insecurity and often a trigger for extra support, where available.)[3] By 2010, the number of children orphaned by AIDS is projected to reach 25 million.[4] Most of these children—20 million of them—will live in sub-Saharan Africa.[5]

LINKS pp. 68, 106

If HIV/AIDS were not boosting mortality rates for adults, the number of children who are orphans would be declining due to overall improvements in human well-being. Unfortunately, at the moment the opposite is happening. In 2001, AIDS orphans accounted for 12.4 percent of all orphans; by 2010, nearly a quarter of orphans will be children who have lost one or both parents to this disease.[6] (See Table 1.)

In 2001, 12 percent of all children in sub-Saharan Africa were orphans, compared with 6.5 percent in Asia and 5 percent in Latin America and the Caribbean.[7] The greater burden of HIV/AIDS in sub-Saharan Africa accounts for much of this difference between regions; without AIDS as a factor, the figure in sub-Saharan Africa would be 8 percent.[8] In Zimbabwe, where an estimated 34 percent of adults are HIV-positive, more than three quarters of the orphans have lost a parent to AIDS.[9] In seven other African nations, orphans due to AIDS account for more than half of the total.[10]

Because it takes about 10 years before an HIV infection leads to an AIDS-related death, given current treatment options and availability in many poor nations, the number of children orphaned by AIDS is expected to continue rising over at least the next decade in countries where HIV is widespread.[11] Botswana and Zimbabwe will be the hardest hit by 2010, with orphans due to AIDS accounting for nearly 90 percent of all children who have lost a parent; in Lesotho, Namibia, Swaziland, and Zambia, the figure is expected to top three quarters.[12]

Even in countries where HIV prevalence has been curbed in recent years, the number of orphans remains high. In Uganda, where adult HIV prevalence declined from 14 percent in the late 1980s to 5 percent in 2001, some 884,000 children have been orphaned by AIDS—one of the largest totals in the world.[13] Although this number is beginning to decline, in 2010 Uganda will still have to care for over a half-million children orphaned by AIDS.[14]

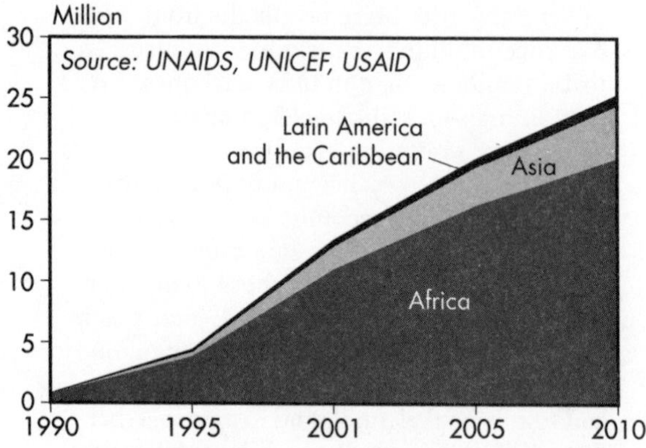

Figure 1: Orphans Due to AIDS, by Region, 1990–2001, with Projections to 2010

In Asia, the number of orphans due to AIDS, 1.8 million, was small compared with the total orphan count of 65.5 million in 2001.[15] Less than 3 percent of orphans in this region lost a parent to AIDS, compared with nearly a third in Africa, reflecting Asia's large populations and low HIV prevalence levels.[16] Yet with growing epidemics in India, China, and Indonesia, even small increases in HIV prevalence can translate into large numbers of AIDS deaths.[17] In fact, projections for Asia indicate that by 2010 orphans due to AIDS will number 4.3 million, accounting for 7.5 percent of all orphans.[18]

Although all three developing regions will

experience an increase in orphans due to AIDS, both in absolute number and as a proportion of all orphans, Africa is distinct in one respect. The total orphan counts in Africa will be higher in 2010 than they are now by about 8 million.[19] In Asia and in Latin America and the Caribbean, in contrast, the total number of orphans is projected to decline slightly because the increase in the number of children orphaned by AIDS will be offset by a decline in the number of orphans from other causes.[20]

One of the most serious effects of AIDS is the increasing number of children who are losing both parents—called "double orphans" by the United Nations. It is likely that one parent who is infected with HIV will pass on the virus to the other. In turn, the eventual death of both parents due to AIDS within a short span of time is also likely. As a result, the number of double orphans who have lost at least one parent to AIDS is expected to increase from 3.8 million in 2001 to 6.9 million in 2010.[21]

All orphans, whether they have lost one parent or two, face tremendous physical and emotional insecurity. The death of a parent can mean greater responsibility to care for siblings, tend to fields, or earn an income. Orphans may face malnutrition and lose access to basic health care, clothing, housing, and education. Young girls are the first to get pulled out of school. UNICEF found that in 20 countries in sub-Saharan Africa, orphans were less likely than other children to be in school and more likely to be working more than 40 hours a week.[22]

Children who lose parents to AIDS can also face stigma and discrimination, including physical abuse and isolation. Without a foster family to care for them, their only option may be prostitution and petty crime while living on the streets. A study in Zimbabwe found that half of all street children are orphans, the majority due to AIDS.[23] Sexual exploitation and drug use heighten the risk that orphaned street children will contract the same virus that their parents succumbed to.

In sub-Saharan Africa, HIV/AIDS is also placing an immense burden on extended families, which often take in orphaned relatives.

Table 1: Orphans Due to AIDS and Other Causes, 1990–2001, with Projections to 2010

Year	Orphans Due to AIDS	Orphans Due to All Other Causes
	(thousand)	(thousand)
1990	903	104,855
1995	4,523	101,923
2001	13,440	94,524
2005	20,106	88,230
2010	25,296	81,469

Source: UNAIDS, UNICEF, and USAID.

Family structures are changing rapidly, with households being headed by single parents, grandparents, other relatives, or children themselves. While people try to cope by increasing income-generating activities, it is often difficult to meet needs, especially in households headed by poor women or elderly grandparents.[24]

Orphanages are not considered an appropriate or effective solution for providing care for orphans. First, the costs associated with building and staffing orphanages are too high for most AIDS-affected countries. Second, child advocacy organizations find that care provided in an institutional setting does not necessarily meet the needs of children, who require personal attention and broader social interactions.

In developing countries, where family and community members are the first to provide care for orphans, direct assistance should be provided to foster families. Policies that strengthen family- and community-based care through day-care centers, support groups, and skills training are also needed. Ultimately, the resources and services available to the larger community and the children themselves will determine how effectively this emerging crisis is managed.[25]

Military and Governance Features

U.S. Airforce photo by Staff Sgt. Greg L. Davis

Corruption Thwarts Development
International Criminal Court Starts Work
Military Expenditures on the Rise
Resource Wars Plague Developing World

Corruption Thwarts Development

Molly O. Sheehan

Corruption—the misuse of public power for private benefit—is hard to measure because officials who take bribes try to hide such activity. Since Transparency International (TI), a Berlin-based nongovernmental organization, published its first global Corruption Perceptions Index in 1995, however, opinion surveys have become a widely used tool to gauge corruption.

The index combines 15 surveys from nine institutions that ask businesspeople, risk analysts, and residents about corruption among public officials and politicians. In 2002, the index covered 102 countries. Of these, 70 nations scored less than 5 out of a clean score of 10, and 35 scored less than 3.[1] (See Table 1.)

LINKS pp. 44, 118, 120

The methodology of this index is evolving, making year-to-year comparisons difficult, but TI does point to a few countries where the perception of corruption seems to be changing.[2] South Korea improved its score between 2001 and 2002, after an anti-corruption law established a commission to investigate high-ranking officials and set fines for bribery of up to $40,000, jail terms of up to 10 years, and a ban on subsequent employment of 5 years.[3] In contrast, Argentina was perceived as being more corrupt in 2002—its economic crisis invited new scrutiny of government spending at the same time that investigations into abuses by former President Carlos Menem were under way.[4]

Corruption erodes people's trust in government. In 1999, two thirds of 57,000 people polled in 60 countries by Gallup's International Millennium Survey believed that their country was not governed by the will of the people.[5] Similarly, the Open Society Institute found that three fourths of citizens in Central and Eastern Europe believed that most or all of their public officials were corrupt, while only 4 out of 10 children surveyed in Europe and Central Asia saw voting in elections as an effective way to improve conditions in their country.[6]

Corruption also appears to sap economic growth. In a path-breaking 1995 paper, economist Paulo Mauro showed that highly corrupt nations have a smaller share of their gross domestic product going into investment.[7] Corruption raises the cost of business, deterring would-be investors.[8] A study of transition economies in Eastern Europe and Central Asia found that gross domestic investment averaged 20 percent less in countries with high corruption compared with countries with medium levels of corruption.[9]

When bribes mean more than votes, a government fails its citizens, as money that could be used to provide needed public services is diverted to private bank accounts. A parliamentary committee in the Philippines calculated in 2002 that corruption costs that government some $1.9 billion annually—twice the size of the national education budget.[10] The World Bank estimates the cost of corruption in Colombia at $2.6 billion a year.[11]

Further, corruption skews public spending toward the sectors where bribing is easier.[12] Studies show that corruption shifts spending away from education, health, and maintenance of existing infrastructure and toward large public works construction and buildup of the military.[13] Indeed, surveying corporate executives, bank officials, and law firms in 15 emerging market economies in 2002, TI found that public works was the sector in which bribes were most often demanded, followed by defense.[14]

At the local level, petty bribes solicited by officials from citizens act as a regressive tax that falls most heavily on the poor. Urban Kenyans polled by the Kenya chapter of TI in early 2001, for example, reported paying some $104 in bribes each month, on an average monthly income of only $331.[15]

Corruption also corresponds to environmental harm. Researchers at Yale University's Center for Environmental Law and Policy have designed an Environmental Sustainability Index that ranks nations by environmental performance. Of 67 quality-of-life variables included in the index, corruption was the one most highly correlated with poor environmental quality.[16] One explanation for this link could be that officials in nations with high levels of corruption take bribes in return for not enforcing environmental laws.

Deforestation spurred by corruption is well

documented, for instance. In Indonesia, a recent study found that many of the logging concessions, covering more than half of the nation's total forest area, were awarded by former President Suharto to relatives and political allies, that at least 16 million hectares of natural forest were approved for conversion to plantations, in direct contradiction of existing laws, and that corrupt officials allowed illegal logging that accounted for some 65 percent of total supply in 2000.[17]

Public officials have also used concessions for mining and fuel extraction to liquidate a nation's resources without passing the revenue on to citizens. In oil-rich Nigeria and Angola, public officials have used oil money for arms and for personal gain.[18] In July 2002, the family of Nigeria's former dictator Sani Abacha agreed to return some $1.2 billion that he took from Nigeria's central bank.[19]

Construction of public works is another area in which corruption has the potential to harm the environment. In Japan, unnecessary and environmentally damaging bridges, dams, and roads have been built as a result of unethical ties between the construction industry and lawmakers.[20] The president of the upper house of Japan's Diet resigned in April 2002 after allegations that his aide took a kickback from a construction company on a public works project.[21]

In a landmark case involving the Lesotho Highlands Water Project, both the briber and the person who was bribed were found guilty of corruption in 2002. A court in Lesotho fined the Canadian company that built a dam $2.2 million for bribing the chief executive of the project, who was sentenced to 18 years in jail; this was the first time a developing nation's court convicted an international company for paying bribes.[22]

The Lesotho case reflects mounting international pressure to combat corruption. Since 1999, the World Bank has barred from development projects companies that are involved in corruption.[23] A 1997 Anti-Bribery Convention by the Organisation for Economic Co-operation and Development (OECD) criminalizes the bribery of foreign public officials, primarily targeting companies from industrial nations that pay bribes for contracts in the developing world.[24] The Asian Development Bank and OECD launched an anti-corruption initiative in the Asia Pacific region in 2001 that committed nations to developing anti-corruption action plans.[25] The Organization of American States has begun to implement an Inter-American Convention against Corruption.[26] And in 2002, the United Nations began negotiating a global treaty on corruption.[27]

All these new initiatives recognize that "it takes two to tango" in corruption: a bribe payer and a bribe taker.[28] But most companies have yet to fear prosecution for paying bribes. A recent survey of managers of major firms operating in developing countries found that only 19 percent knew something about the 1997 OECD anti-bribery treaty.[29] Peter Eigen, TI's chairman, notes: "Only a level playing field—a world in which honest companies know that bribery doesn't pay and that unscrupulous competitors will be punished—will bring about a lasting change in the behavior of international business."[30] To help level the playing field, his organization is working with companies such as BP, Shell, Tata, and General Electric to develop business principles for countering bribery.[31]

Table 1: Nations Perceived by Business People and Risk Analysts as Most Corrupt of 102 Surveyed, 2002

Corruption Index Score[1]	Countries
1.0–1.9	Bangladesh, Nigeria, Angola, Madagascar, Paraguay, Indonesia, Kenya
2.0–2.3	Azerbaijan, Moldova, Uganda, Bolivia, Cameroon, Ecuador, Haiti, Kazakhstan
2.4–2.6	Georgia, Ukraine, Viet Nam, Albania, Guatemala, Nicaragua, Venezuela, Pakistan, Philippines, Romania, Zambia
2.7–2.9	Côte d'Ivoire, Honduras, India, Russia, Tanzania, Zimbabwe, Argentina, Malawi, Uzbekistan

[1] Index ranges from 0 (most corrupt) to 10 (least corrupt).

Source: Transparency International.

International Criminal Court Starts Work

Arunima Dhar

On July 1, 2002, the Rome Statute of the International Criminal Court (ICC) entered into force, creating the first permanent and independent court capable of investigating the most serious violations of international humanitarian law—genocide, war crimes, crimes against humanity, and, once it is defined, aggression.[1] (See Table 1 for an overview of the ICC.) Countries that ratify the statute agree either to prosecute individuals accused of such crimes under their own laws or to surrender them to the Court for trial.[2]

LINKS pp. 74, 76

The Court's jurisdiction is not retroactive: it can only address crimes that have occurred on or after July 1, 2002.[3] The ICC will act as a court of last resort, when national courts are unable or unwilling to act. As adopted, the ICC statute also asserts jurisdiction over citizens of a non-party state by virtue of the "universal jurisdiction" exercised by the ICC.[4] But the country where the crimes are alleged to have occurred has to have ratified the statute or at least accept the Court's jurisdiction, even if only on a temporary basis. The U.N. Security Council can also refer a case to the Court.

The U.N. General Assembly first recognized the need for a permanent judicial mechanism to prosecute war criminals and mass murderers in 1948, following the Nuremberg and Tokyo trials after World War II. Until the adoption of the Rome Statute of the International Criminal Court, there was no system to hold individuals criminally liable for genocide or war crimes. The International Court of Justice at the Hague, the principal judicial organ of the United Nations, was designed to deal primarily with disputes between states, and hence it had no jurisdiction over matters involving individual criminal responsibility.[5]

Since World War II, the need for a permanent criminal court has only become stronger. An estimated 170 million people have died in 250 conflicts.[6] From April to July 1994, for example, some 800,000 Rwandans—roughly 10 percent of that country's population—were murdered during ethnic violence between Hutus and Tutsis.[7] An estimated 10,000 ethnic Albanians were killed in Kosovo by the armed forces of the Federal Republic of Yugoslavia from March to early June 1999.[8] On August 30, 1999, East Timor voted in favor of independence from Indonesia; following the vote, militia forces massacred hundreds, and possibly thousands, of East Timorese.[9] And about 300,000 people perished under the rule of Idi Amin, the de facto President of Uganda from 1971 to 1979.[10]

The U.N. Security Council's establishment of the International Criminal Tribunals for the former Yugoslavia and for Rwanda in 1993 and 1994 was perceived as a step in the right direction.[11] The establishment of tribunals has also been pursued in a few other areas. In East Timor, a "hybrid" approach—involving international and local judges—is in operation.[12] In July 2002, the U.N. Security Council approved plans for the launch of a special court to try people for abuses committed during the civil war in Sierra Leone.[13] But these panels suffer from inadequate funding and what critics have called low standards of professionalism.[14] In Cambodia, U.N. officials have been negotiating with the government to set up a tribunal for Khmer Rouge leaders, but ensuring adequate standards has been a major obstacle.[15] And these tribunals are limited in their territorial and temporal jurisdiction. The ICC was created with the goal of deterring future crimes and remedying the deficiencies of ad hoc tribunals.

For nearly a decade, the United States had demonstrated a consistent level of executive and congressional interest in the concept of a permanent International Criminal Court. The Foreign Operations, Export Financing, and Related Programs Appropriations Act of 1991, for instance, required the President and the Judicial Conference of the United States to report to the Congress on the desirability of establishing an International Criminal Court.[16]

But the United States sought to weaken the Rome Statute and ended up being one of seven nations voting against it (along with China, Iraq, Libya, Yemen, Qatar, and Israel). The Statute was adopted in July 1998 by 120 nations voting in favor, with 21 abstaining.[17]

Table 1: Overview of the International Criminal Court

Characteristic	Description
Origin	Legal framework was established at a U.N.-sponsored conference in Rome with representatives from 160 countries. Rome Statute was adopted on 17 July 1998. By the deadline of 31 December 2000, 139 countries had signed the statute; as of 10 February 2003, 89 countries had ratified it.
Crimes dealt with	Genocide (the commission of certain acts "with the intent to destroy in whole or in part, a national, ethnic, racial, or religious group, as such"); crimes against humanity ("committed as part of a widespread or systematic attack directed against any civilian population, with knowledge of the attack, pursuant to or in furtherance of a national or organizational policy"); and currently undefined crimes of aggression (under the ICC's jurisdiction, there was not enough time to reach a definition of "aggression" that was acceptable to all).
Status	The ICC is an independent and permanent court.
Cost and funding	The Court will be funded by assessed contributions from states party to it; by funds provided by the United Nations; and by voluntary contributions from governments, international organizations, individuals, corporations, and other entities.
Referral to ICC	A case may be referred to the ICC by a country member of the Assembly of State Parties, by a country that has chosen to accept the ICC's jurisdiction, by the Security Council (subject to veto), or by the three-judge panel when it authorizes a case initiated by the International Prosecutor.
Penalty and compensation	Consistent with international human rights standards, the ICC has no competence to impose a death penalty. The Court may impose lengthy terms of imprisonment of up to 30 years or life when justified by the gravity of the case. The Court may, in addition, order a forfeiture of proceeds, property, or assets derived from the committed crime.

Source: Rome Statute of the International Criminal Court, Coalition for the International Criminal Court, and Human Rights Watch.

President Clinton signed on to the ICC in the last days of his term in office.[18] Yet his successor launched an aggressive campaign to undermine the court and to remove U.S. nationals from its jurisdiction. On May 6, 2002, the Bush administration withdrew the U.S. signature.[19] The main reason cited for its opposition was the belief that U.S. troops would be subjected to frivolous or politically motivated investigations and that the ICC would develop into a "wide open complaint system."[20]

In August 2002, President Bush signed into law the American Servicemembers' Protection Act of 2002, dubbed by activists as the Hague Invasion Act.[21] The new law prohibits U.S. cooperation with the ICC and authorizes the use of military force to liberate any American or a citizen of a U.S.-allied country being held by the Court. In addition, the law provides for the withdrawal of U.S. military assistance from countries ratifying the ICC treaty and restricts U.S. participation in U.N. peacekeeping operations unless the United States obtains immunity from prosecution by the ICC.[22] Threatening to veto the renewal of a U.N. peacekeeping mission in Bosnia, the United States secured a one-year promise of immunity from prosecution for peacekeepers from the United States and other states that are not member of the ICC.[23]

The United States has also signed impunity agreements that seek to prevent U.S. nationals accused of genocide, crimes against humanity, or war crimes from being surrendered to the International Court. The government has threatened to cut off military aid to any state that is party to the Rome Statute that does not enter into an impunity agreement with the United States. As of November 18, 2002, a total of 15 countries had signed such agreements.[24]

Military Expenditures on the Rise

Michael Renner

According to the Stockholm International Peace Research Institute, world military expenditures amounted to a conservatively estimated $839 billion in 2001, the most recent year for which data are available.[1] This works out to $2.3 billion each day—almost $100 million an hour.[2]

The reduction in military expenditure that began in the fading days of the cold war in 1987 is now definitely a thing of the past. From $847 billion in 1992 (in 1998 dollars), military budgets fell to $719 billion in 1998.[3] Since then, spending has been on the rise again, however, with the total in 2001 equaling $772 billion in 1998 dollars.[4] And with the "war on terrorism" in full swing, spending appears set for substantial further increases, particularly in the United States.[5]

LINKS p. 74

World military spending amounted to $137 per capita in 2001.[6] Such an average figure, however, masks massive imbalances among different countries.[7] More than three quarters of the total is spent by just 15 countries.[8] The United States is now the world's sole military colossus, accounting for 36 percent of all military spending—as much as the next nine biggest spenders combined.[9]

The other nine countries can be grouped into two tiers. The first includes Russia, France, Japan, and the United Kingdom—together accounting for 21 percent of world spending.[10] The second encompasses Germany, China, Saudi Arabia, Italy, and Brazil—with a combined 15 percent share.[11]

The United States and its core allies in Europe and the Asia-Pacific region (the North Atlantic Treaty Organization, Japan, South Korea, and Australia) account for two thirds of world spending.[12] (See Figure 1.) In stark contrast, Russia and China combined spend less than 10 percent.[13] And the group of countries that the U.S. government considers hostile "rogue states" (Cuba, Iran, Iraq, Syria, Libya, and North Korea) barely register on a global scale. Collectively, they spend less than 3 percent of the world total, or one fourteenth the U.S. budget.[14]

Increasingly, the U.S. military—particularly its technological edge and global reach—is without peer.[15] U.S. military R&D, slated to grow from $56.8 billion in fiscal year 2003 to $61.8 billion in 2004, is the fastest-growing component of the American military budget.[16] It alone surpasses the total amount that any other nation devotes to military purposes.

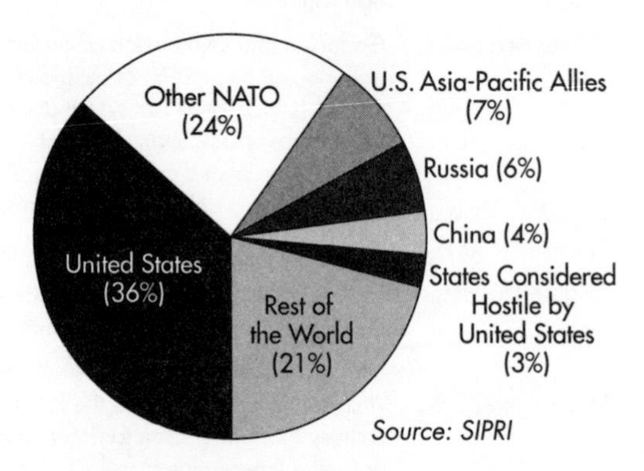

Figure 1: Share of World Military Expenditures, by Country or Group, 2001

The events of 11 September 2001 provided political cover for the largest expansion in the U.S. military budget in 20 years, even though most of the extra money is not slated for the "war on terrorism" as such. President Bush requested an additional $48 billion in fiscal year 2002 funds.[17] Large additional increases are now proposed. The plan submitted to Congress by the Bush administration in February 2003 envisions a military budget of $503 billion in fiscal year 2009 ($414 billion in inflation-adjusted dollars of 2001), compared with $363 billion in 2003.[18] (See Figure 2.) And the administration may ask Congress for even more money.[19]

The Pentagon's fiscal year 2004 weapons procurement request stands at $73 billion.[20] Just under half of that money goes to fund the top 20 weapons systems, in what is little more than a down payment for huge costs over the years. The total program costs of these 20 systems—including jet fighters, combat helicopters, destroyers, attack submarines, and high-tech missiles and munitions—are currently estimated at $750 billion.[21] But such systems have historically suffered from large cost overruns, so a commitment to them may imply even greater expenses in the future.

A swelling military budget reflects the Bush administration's policy of unquestioned military superiority. In a national security strategy statement released in 2002, the U.S. administration asserted that "we must build and maintain our defenses beyond challenge" to dissuade any other powers from even considering a military buildup.[22] And it says that "to forestall or prevent...hostile acts by our adversaries, the United States will, if necessary, act preemptively."[23]

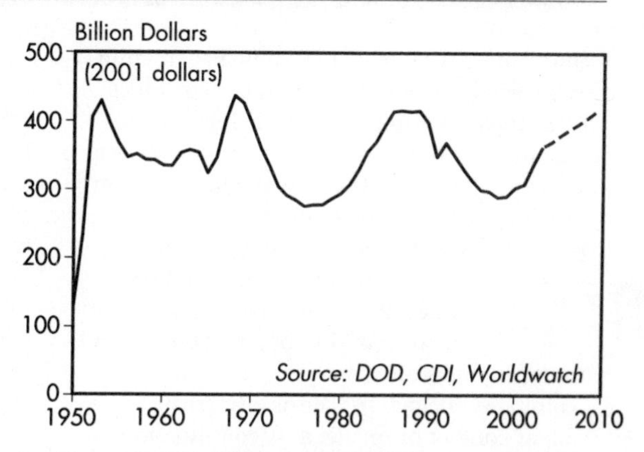

Figure 2: U.S. Military Expenditures, 1950–2003, with Projections to 2009

The confrontation with Iraq is the first concrete application of this policy. A war against Iraq would likely absorb additional tens of billions of dollars. In December 2002, the White House put forward an estimate of war costs of $50–60 billion.[24] But this did not include the possibility of a protracted conflict. Nor did it address the costs associated with an occupation regime or with humanitarian and reconstruction needs. And it did not take into account the possible wider economic repercussions of a war. These could well be in the hundreds of billions of dollars.[25]

The 32 richest nations, with just 15 percent of the world's population, spend by far the most on the military, representing 70 percent of global expenditures in 2001.[26] However, the most militarized countries—with the highest per capita spending—are located in the Middle East.[27] States in that region imported close to $190 billion (in 2001 dollars) worth of weapons from 1990 to 2001; Saudi Arabia and other Persian Gulf states accounted for almost two thirds of that sum.[28]

By contrast, 51 low-income countries, with 41 percent of world population, account for just 7 percent of total military expenditures.[29] Yet this is more than double their current portion of the world's gross economic product, and military spending is a heavy burden for these impoverished and indebted nations. For countries such as Eritrea, Burundi, and Pakistan, military spending equals or surpasses combined public expenditures for health and education.[30]

Some governments are setting different priorities. Brazil decided in January 2003 to put on hold the planned purchase of jet fighters costing $760 million.[31] And the military budget is to be cut 4 percent in a bid to fund an ambitious anti-hunger program.[32]

The renewed growth in world military expenditure has troubling implications for many of the world's unmet needs in the fields of health, education, general human well-being, and environmental protection. Estimates suggest, for instance, that the prevention of soil erosion worldwide would require something on the order of $24 billion annually; the elimination of starvation and malnutrition, $19 billion; reproductive health for all women, $12 billion; safe, clean drinking water, $10 billion; prevention of acid rain, $8 billion; and elimination of illiteracy, $5 billion.[33] Although these are substantial sums, they pale in comparison with the funds being made available for military purposes.

Half a century after they were first spoken, U.S. President Dwight Eisenhower's warning words still ring true today: "Every gun that is fired, every warship launched, every rocket fired signifies, in the final sense, a theft from those who hunger and are not fed, those who are cold and are not clothed."[34]

Resource Wars Plague Developing World

Michael Renner

Abundant natural resources—such as oil, minerals, metals, diamonds and other gemstones, drug crops, and timber—have helped fuel a large number of armed conflicts in developing countries. Resource wealth plays an important role in the outbreak of conflict and tends to make conflicts last longer, although it has a more varied influence on their intensity. Altogether, in about a quarter of the roughly 50 wars and armed conflicts of recent years, resource exploitation helped trigger or exacerbate violent conflict or financed its continuation.[1]

LINKS pp. 74, 76, 98, 102, 114

In those cases, natural resource wealth has turned out to be a curse, triggering a torrent of arms trafficking, human rights violations, humanitarian disasters, and environmental destruction. A rough, conservative estimate suggests that more than 5 million people were killed in resource-related conflicts during the 1990s.[2] (At least 2.5 million people died in the conflict in the Democratic Republic of the Congo alone).[3] In addition, close to 6 million fled to neighboring countries, and anywhere from 11–15 million people were displaced inside the borders of their home countries.[4]

The money derived from resource exploitation in war zones has secured an ample supply of arms and military equipment for armed factions and has served to enrich a handful of people—warlords, corrupt government officials, arms merchants, mercenaries, and unscrupulous corporate leaders.[5] (See Table 1.) But critical human needs have been trampled in the process. In oil- and diamond-rich Angola, for instance, almost 30 percent of children die before the age of six.[6] Nearly half of all Angolan children are underweight, and a third of school-age children have no school to go to.[7] Unsafe drinking water, a pervasive lack of health services, and food shortages have limited Angolans' life expectancy to 47 years.[8]

In places like Angola, Afghanistan, Cambodia, and Colombia, the pillaging of resources allows wars to continue that were initially driven by grievances or liberation and ideological struggles.[9] Elsewhere, such as in Sierra Leone or the Democratic Republic of the Congo, nature's bounty attracts predatory groups that initiate violence as a means of establishing control over resource deposits.[10] Finally, resource extraction can itself be the source of conflict where the economic benefits accrue to foreign companies and local elites, while the local population shoulders an array of the burdens. This has led to violent conflict in places like Nigeria's Niger Delta, Papua New Guinea's Bougainville island, and several provinces in Indonesia.[11]

Violent struggles arising out of a context of contested resource wealth join a host of conflicts that emerge from situations of resource scarcity—overuse and depletion—and are exacerbated by the social and economic repercussions of environmental degradation. Where resource wealth is a factor in conflicts, it is primarily nonrenewable resources such as fuels and minerals that are at issue. Where resource scarcity is a factor, on the other hand, it is the degradation of arable land, the depletion of water for irrigation and drinking, and the decimation of forests that are focal points.[12]

Combatants have relied on a variety of means to secure the natural resources that finance their military activities. They use extreme violence to establish undisputed control, intimidate local populations, or de-populate resource-rich areas altogether. They pillage existing stocks, coerce large numbers of civilians into mining and logging operations, or put some of their own combatants to work. They "tax" loggers and miners or otherwise extract ransom before allowing the passage of commodities to their intended markets. They contract with unscrupulous companies to extract, smuggle, and market the resources.[13]

The countries with resource-related conflicts suffer from a number of debilitating economic and political conditions. Overly dependent on natural resources, they fail to diversify their economies, stimulate innovation, or invest adequately in critical social areas or public infrastructure.[14] Resource royalties help political leaders maintain power, even in the absence of popular legitimacy, by funding a system of patronage.[15] These governments also spend a

Table 1: Estimated Revenues from Conflict Resources, Selected Cases

Combatant	Resource	Period	Estimated Revenue
UNITA rebels (Angola)	diamonds	1992–2001	$4–4.2 billion total
RUF rebels (Sierra Leone)	diamonds	1990s	$25–125 million a year
Liberia (government)	timber	Late 1990s	$100–187 million a year
Sudan (government)	oil	Since 1999	$400 million a year
Rwandan army	coltan (from Congo)	1999–2000	$250 million total
Afghanistan (Taliban, Northern Alliance)	opium, lapis lazuli, emeralds	Mid-1990s–2001	$90–100 million a year
Cambodia (government, Khmer Rouge)	timber	Mid-1990s	$220–390 million a year
Myanmar (government)	timber	1990s	$112 million a year
FARC rebels (Colombia)	cocaine	Late 1990s	$140 million a year

Source: Compiled from Renner, *The Anatomy of Resource Wars.*

high portion of state income on internal security to suppress challenges to their power.[16] Ruthless predatory groups have emerged, intent on seizing control of a prized resource that represents one of the few tickets to wealth and power. Violent tactics are facilitated by the massive proliferation and easy availability of small arms and light weapons.[17]

Ending these kinds of conflicts and the associated pillage is not easy. In the Congo, foreign forces have withdrawn, yet fighting among various armed factions continues, and elaborate illegal networks have emerged that continue to exploit natural resources for the benefit of a handful of Congolese, Zimbabwean, Ugandan, and Rwandan elites.[18]

The enormous expansion of global trade and financial networks has made access to key markets relatively easy for warring groups. They have had little difficulty in establishing international smuggling networks and sidestepping existing international embargoes, given a degree of complicity among certain companies and lax customs controls.[19]

It is at least becoming a bit more difficult for "conflict resources" to be sold on world markets. In the diamond industry, this is due to national certification schemes and efforts to negotiate a standardized global certification scheme. But the resulting set of rules still suffers from numerous shortcomings, including reliance on voluntary measures and a lack of independent monitoring.[20]

Natural resources will continue to fuel deadly conflicts as long as consumer societies import materials with little regard for their origin or the conditions under which they were produced. Some civil society groups have sought to increase consumer awareness and to compel companies—some of them major corporations—to do business more ethically through investigative reports and by "naming and shaming" specific corporations.[21]

Promoting democratization, justice, and greater respect for human rights are key tasks, along with efforts to reduce the impunity with which some governments and rebel groups engage in extreme violence. Another challenge is to diversify economies away from a strong dependence on primary commodities. A more diversified economy, and investments in human development, would lessen the likelihood that natural resources become pawns in a struggle among ruthless contenders for wealth and power.

Notes

GRAIN PRODUCTION DROPS (pages 28–29)

1. U.N. Food and Agriculture Organization (FAO), *FAOSTAT Statistical Database*, at <apps.fao.org>, updated 9 January 2003; idem, *Food Outlook* (Rome: December 2002), p. 1; U.S. Department of Agriculture (USDA), *Grain: World Markets and Trade* (Washington DC: October 2002).
2. FAO, *FAOSTAT*, op. cit. note 1; idem, *Food Outlook*, op. cit. note 1, p. 1.
3. FAO, *FAOSTAT*, op. cit. note 1.
4. Ibid.
5. FAO, *Food Outlook*, op. cit. note 1, p. 1; corn figures from Paul Racionzer, economist, Global Information and Early Warning Service, Commodities and Trade Division, FAO, e-mail to author, 17 January 2003.
6. FAO, *FAOSTAT*, op. cit. note 1.
7. FAO, *Food Outlook*, op. cit. note 1, pp. 36–37.
8. USDA, op. cit. note 1, p. 1.
9. FAO, *Food Outlook*, op. cit. note 1, pp. 36–37; idem, *Food Outlook* (Rome: October 2002), pp. 28–29.
10. FAO, *Food Outlook*, op. cit. note 1, pp. 36–37.
11. Ibid.; USDA, op. cit. note 1, p. 7.
12. FAO, *Food Outlook*, op. cit. note 1, pp. 36–37; USDA, op. cit. note 1, p. 1.
13. FAO, *FAOSTAT*, op. cit. note 1.
14. Per capita figures calculated from FAO, *FAOSTAT*, op. cit. note 1, and from U.S. Bureau of the Census, *International Data Base*, electronic database, Suitland, MD, updated 10 October 2002.
15. Ibid.
16. FAO, *The State of Food and Agriculture* (Rome: 2002), pp. 3–5.
17. Ibid.; 40 million from James T. Morris, "Crisis in Africa: The Political Dimensions of Hunger," Statement to the U.N. Security Council, New York, 3 December 2002.
18. Morris, op. cit. note 17; FAO, *Food Outlook*, op. cit. note 1, p. 7; idem, op. cit. note 9, p. 7.
19. FAO, *FAOSTAT*, op. cit. note 1; idem, *Food Outlook*, op. cit. note 1, pp. 1–5.
20. FAO, *FAOSTAT*, op. cit. note 1; idem, *Food Outlook*, op. cit. note 1, pp. 19–20.
21. FAO, *FAOSTAT*, op. cit. note 1; idem, *Food Outlook*, op. cit. note 1, pp. 19–20.

MEAT PRODUCTION AND CONSUMPTION GROW (pages 30–31)

1. U.N. Food and Agriculture Organization (FAO), *FAOSTAT Statistical Database*, at <apps.fao.org>, updated 9 January 2003; idem, "Meat and Meat Products," *Food Outlook No. 4*, October 2002, p. 11.
2. FAO, *FAOSTAT Statistical Database*, op. cit. note 1.
3. Ibid.
4. Ibid.
5. Christopher Delgado et al., *Livestock to 2020: The Next Food Revolution* (Washington, DC: International Food Policy Research Institute, 1999), p. 1.
6. FAO, *FAOSTAT Statistical Database*, op. cit. note 1.
7. Ibid.
8. Ibid.
9. Ibid.
10. Ibid.
11. Ibid.
12. Ibid.
13. Cees de Haan et al., "Livestock & the Environment: Finding a Balance," Report of a Study Coordinated by FAO, U.S. Agency for International Development, and World Bank (Brussels: 1997), p. 53.
14. U.S. Department of Agriculture (USDA), *Production, Supply, and Distribution*, electronic database, Washington, DC, updated 27 November 2002.
15. Vaclav Smil, Department of Geography, University of Manitoba, Canada, discussion with Brian Halweil, October 2002.
16. FAO, *FAOSTAT Statistical Database*, op. cit. note 1.
17. Ibid.
18. De Haan et al., op. cit. note 13, pp. 54–55.
19. Ibid., pp. 17–18.
20. USDA Food Safety and Inspection Service, at <www.fsis.usda.gov/index.htm>, viewed 2 August

2002; Greg Winter, "Beef Processor No Stranger to Troubles," *New York Times*, 20 July 2002.
21. FAO, *FAOSTAT Statistical Database*, op. cit. note 1.
22. Ibid.
23. Kim Severson, "Grassroots Revolution—Will the New Beef Put Corn-raised Cattle Out to Pasture?" *San Francisco Chronicle*, 19 June 2002.

FOSSIL FUEL USE UP (pages 34–35)

1. Calculated with data and information from David Fridley, Lawrence Berkeley Laboratory, e-mail to Anand Rao, Worldwatch Institute, 31 January 2003, from U.S. Department of Energy (DOE), Energy Information Administration (EIA), *Monthly Energy Review*, January 2003, from Terence H. Thorn, International Gas Union, "Actual Developments in the World Natural Gas Industry," January 2003, at <www.igu.org/index.asp?p_link=links/gas_oil.asp>, viewed 3 February 2003, from International Energy Agency (IEA), *Oil Market Report*, 17 January 2003, and from BP, *Statistical Review of World Energy 2002* (London: June 2002).
2. Based on data from BP, op. cit. note 1.
3. Increase since 1950 based on sources cited in note 1; share of total global energy use calculated by Worldwatch with 2000 data from IEA, *World Energy Outlook 2002* (Paris: 2002), pp. 410–11.
4. Increase in global oil consumption from IEA, op. cit. note 1.
5. U.S. share of world oil demand from BP, op. cit. note 1, p. 10; increase in oil use from IEA, op. cit. note 1.
6. IEA, op. cit. note 1.
7. Ibid.
8. Ibid.
9. Based on sources cited in note 1; increase relative to 2001 calculated with data from BP, op. cit. note 1.
10. Calculated by Worldwatch with data through October 2002 from DOE, op. cit. note 1, p. 90.
11. China's share of global consumption from BP, op. cit. note 1, p. 33; increase over 2001 calculated with data from BP and 2002 additions provided by Fridley, op. cit. note 1.
12. Banning of coal in certain regions from "Cheap Coal a Hurdle to China Natgas Growth—Expert," *Reuters*, 30 January 2003; increasing output from "China's Surging Coal Production: Fact or Fiction?" *China Coal Report* (Barlow Jonker Pty, Ltd.), 21 August 2002.
13. Growth rate from Thorn, op. cit. note 1; total based on ibid. and on BP, op. cit. note 1, p. 25.
14. Share of global gas use from BP, op. cit. note 1, p. 25; decline relative to 2001 derived from data in DOE, op. cit. note 1, p. 73.
15. Gross consumption by countries in the Organisation for Economic Co-operation and Development; IEA, *Monthly Natural Gas Survey*, November 2002, pp. 2–6.
16. Based on data from BP, op. cit. note 1.
17. DOE, EIA, *International Energy Outlook 2002* (Washington, DC: March 2002), p. 44.
18. Thorn, op. cit. note 1.
19. IEA, op. cit. note 3, p. 26.
20. Ibid., pp. 27–28.
21. Ibid., p. 32.
22. IEA assumptions from ibid., pp. 46–54; peaking of world oil production from "Analysts Claim Early Peak in World Oil Demand," *Oil & Gas Journal Online*, 12 August 2002, at <ogj.articles.printthis.clickability.com/pt/cpt?expire=&fb=Y&urlID=3854003&action=cpt&partnerID=1400>, viewed 12 February 2003.

NUCLEAR POWER RISES (pages 36–37)

1. Installed nuclear capacity is defined as reactors connected to the grid as of 31 December 2002 and is based on Worldwatch Institute database compiled from statistics from the International Atomic Energy Agency and press reports primarily from *Associated Press*, *Reuters*, *World Nuclear Association (WNA) News Briefing*, and Web sites.
2. Worldwatch Institute database, op. cit. note 1.
3. Ibid.; an additional seven reactors in Canada that had been mothballed are in varying stages of being restarted, though no successful restart is certain.
4. Worldwatch Institute database, op. cit. note 1.
5. Ibid.
6. World Nuclear Association, "Nuclear Power in China," November 2002.
7. "India to Increase its Nuclear Power Generation 10-Fold by 2020," *Press Trust of India*, 28 October 2002.
8. "British Energy Reported Higher than Expected Losses," *WNA News Briefing*, 11–17 December 2002.
9. "British Energy H1 Worse than Expected, Amergen Talks Continue," *AFX News Limited*, 12 December 2002, from PowerMarketers.com, at <www.powermarketers.com>, viewed 16 December 2002.
10. "Belgium: The Lower House of the Belgium Parliament Has Voted," *WNA News Briefing*, 4–10 December 2002.

11. International Atomic Energy Agency (IAEA), "Latest News Related to PRIS and the Status of Nuclear Power Plants," at <www.iaea.org>, viewed 20 January 2003.
12. "Romania: The Government Has Signed An Agreement," *WNA News Briefing*, 11–17 December 2002.
13. Vladimir Slivyak, Co-Chairman, ECODEFENSE! and Director, Anti-Nuclear Campaign, Socio-Economic Union, Nizhegorodskaya, e-mail to author, 10 January 2003.
14. "Exelon Deals SA Nuclear Energy Plan a Major Blow," *Business Day/All Africa Global Media*, 17 April 2002.
15. Will McNamara, "Early Site Permit Process Moves Three Companies Closer to Building a New Nuclear Reactor," *Scientech IssueAlert*, 9 May 2002.
16. Quote from "Japan Nuclear Industry Stunned by Tepco Cover-up," *Platts News*, at <www.platts.com/archives/91183.html>, viewed 4 September 2002; Chihiro Kamisawa and Satoshi Fujino, "Revelation of Endless N-damage Cover-ups: The 'TEPCO Scandal' and the Adverse Trend of Easing Inspection Standards," *Nuke Info Tokyo*, November/December 2002.
17. Kamisawa and Fujino, op. cit. note 13.
18. "Permission Withdrawn for Building Two Tsuruga Reactors," *Platts News*, at <www.platts.com/archives/91293.html>, viewed 10 September 2002; "Prefectural Governments of Niigata and Fukushima Completely Retract Their Prior Consent to the Plu-Thermal Program," *Nuke Info Tokyo*, November/December 2002.
19. "Kepco Set to Build Eight New Korean Nuclear Plants," *Power Engineering International*, at <pei.articles.com/Articles/Article_Disply.cfm?ARTICLE_ID=142945>, viewed 13 May 2002.
20. "U.S.: Iran Working on Nuclear Weapons," *CNN.com*, 13 December 2002.
21. "Iran Says Committed to Nuclear Power Programme," *Reuters*, 14 December 2002.
22. "N. Korea Reactor Work, Oil Supply Unchanged—Official," *Reuters*, 21 October 2002.

WIND POWER'S RAPID GROWTH CONTINUES (pages 38–39)

1. Worldwatch Institute preliminary estimate based on figures from Birger T. Madsen, BTM Consult, e-mails to author, 23 January and 26 February 2003, on European Wind Energy Association (EWEA), "European Wind Industry; Another Record Year," press release (Brussels: 6 February 2003), on American Wind Energy Association (AWEA), "U.S. Wind Industry Turns in Another Solid Year of Growth," press release (Washington, DC: 23 January 2003), and on Bundesverband WindEnergie e.V. (BWE), "Erneutes Rekordjahr für die Windbranche—12.000 Megawatt überschritten," press release (Berlin: 22 January 2003). Historical data from BTM Consult, *International Wind Energy Development: World Market Update 2001* (Ringkøbing, Denmark: March 2002); historical numbers have been adjusted to incorporate more recent BTM Consult estimates.
2. Worldwatch Institute estimate based on sources in note 1.
3. Figure of 35 million from EWEA, "Wind Energy—Clean Power for Generations," brochure (Brussels: 2002).
4. Addition in 2002 is Worldwatch Institute preliminary estimate based on sources in note 1.
5. Additions in 2002 and year-end capacity from AWEA, op. cit. note 1; capacity in 2001 from AWEA, "U.S. Wind Industry Ends Most Productive Year, More than Doubling Previous Record for New Installations," press release (Washington, DC: 15 January 2002).
6. AWEA, op. cit. note 1.
7. EWEA, op. cit. note 1.
8. Worldwatch Institute estimates based on sources in note 1.
9. Ibid.
10. BWE, op. cit. note 1.
11. Reductions are relative to 1990 levels and assume that the European Union will agree to reductions of 30 percent overall; German Environment Minister Jürgen Trittin, "The Success Story of Climate Protection in Germany," speech, New Delhi, India, 1 November 2002.
12. According to Spain's Association of Renewable Energy Producers, cited in "Spanish Wind Power Capacity rises 44 pct in 2002," *Reuters*, 4 February 2003.
13. Capacity figures from EWEA, op. cit. note 1; generating potential from Charles P. Wallace, "Is It a Breeze?" *TIME Europe*, 22 July 2002.
14. Madsen, op. cit. note 1.
15. Matthew Jones, "UK to Accelerate Pace for Renewable Energy Target," *Reuters*, 14 February 2002.
16. Ibid.
17. Capacity additions from EWEA, op. cit. note 1; total calculated with 2001 data from BTM Consult, op. cit. note 1, p. 10.

18. Madsen, op. cit. note 1; BTM Consult, op. cit. note 1, p. 10.
19. Madsen, op. cit. note 1.
20. Madsen, op. cit. note 1; BTM Consult, op. cit. note 1, p. 10.
21. China capacity from Madsen, op. cit. note 1; expected growth from Tim Sharp, "Asian Wind Market Shows Eastern Promise," *Platts Renewable Energy Report*, August 2002, pp. 37–39.
22. Sharp, op. cit. note 21.
23. Worldwatch estimate based on Andreas Wagner, GE Wind Energy and EWEA, e-mail to author, 18 September 2002, and on EWEA, Forum for Energy and Development, and Greenpeace, *Wind Force 10* (London: 1999).
24. EWEA, op. cit. note 3.
25. "Wind Turbine Report Available," SolarAccess.com, Norwalk, CT, 23 January 2003, at <www.solaraccess.com/news/story?storyid=3495&p=1>, viewed 31 January 2003.
26. Merrill Lynch cited in Wallace, op. cit. note 13.

CARBON EMISSIONS AND TEMPERATURE CLIMB (pages 40–41)

1. James Hansen et al., NASA's Goddard Institute for Space Studies, New York, Surface Air Temperature Analyses, "Global Land-Ocean Temperature Index in .01 C," at <www.giss.nasa.gov/data/update/gistemp/GLB.Ts+dSST.txt>, viewed 13 January 2003.
2. National Oceanic and Atmospheric Administration (NOAA)'s National Climatic Data Center, Asheville, NC, "Climate of 2002 Annual Review," 23 January 2003, at <www.ncdc.noaa.gov/oa/climate/research/2002/ann/ann02.html>, viewed 24 January 2003; Jean Palutikof, University of East Anglia's Climatic Research Unit, Norwich, UK, at <www.cru.uea.ac.uk/cru/info/warming/>, viewed 24 January 2003.
3. J. T. Houghton et al., eds., *Climate Change 2001: The Scientific Basis*, Contribution of Working Group I to the Third Assessment Report of the Intergovernmental Panel on Climate Change (Cambridge, UK: Cambridge University Press, 2001).
4. Carbon emissions calculated with data from Gregg Marland, Tom Boden, and Robert J. Andres, "Global CO_2 Emissions from Fossil-Fuel Burning, Cement Manufacture, and Gas Flaring: 1751–1999," Carbon Dioxide Information Analysis Center, Oak Ridge National Laboratory and University of North Dakota, 9 December 2002, at <cdiac.esd.ornl.gov/ftp/ndp030/global99.ems>, viewed 3 February 2003, from British Petroleum (BP), *Statistical Review of World Energy* (London: June 2002), from David Fridley Lawrence Berkeley Laboratory, e-mail to Anand Rao, Worldwatch Institute, 31 January 2003, from U.S. Department of Energy, Energy Information Administration, *Monthly Energy Review*, January 2003, from Terence H. Thorn, International Gas Union, "Actual Developments in the World Natural Gas Industry," January 2003, at <www.igu.org/index.asp?p_link=links/gas_oil.asp>, viewed 3 February 2003, and from International Energy Agency, *Oil Market Report*, 17 January 2003; carbon dioxide concentrations at Mauna Loa Observatory, Hawaii, from Timothy Whorf, Scripps Institution of Oceanography, La Jolla, CA, discussion with Janet Sawin, 30 January 2003.
5. Whorf, op. cit. note 4.
6. Houghton et al., op. cit. note 3.
7. Ibid.
8. Richard A. Anthes, *Meteorology*, 7th ed. (Upper Saddle River, NJ: Prentice Hall, 1997), pp. 173–76.
9. David A. Ross, *Introduction to Oceanography* (New York: HarperCollins College Publishers, 1995), p. 323.
10. Ibid.; Whorf, op. cit. note 4.
11. World Meteorological Organization, "El-Niño Outlook," Geneva, 29 August 2002; World Meteorological Organization, "WMO Statement on the Status of the Global Climate in 2002: Global Surface Temperatures Second Warmest on Record, Return to El Niño Conditions," press release (Geneva: 17 December 2002).
12. Hadley Centre for Climate Prediction and Research, Met Office, "2002: One of the Warmest Years on Record in Britain, Second Warmest Year Globally," press release (Bracknell, UK: 17 December 2002); Andrew Revkin, "Temperatures Are Likely to Go From Warm to Warmer," *New York Times*, 31 December 2002.
13. Camille Parmesan and Gary Yohe, "A Globally Coherent Fingerprint of Climate Change Impacts Across Natural Systems," *Nature*, 2 January 2003, pp. 37–42.
14. Houghton et al., op. cit. note 3, Chapter 2.
15. Ibid., Chapter 11.
16. James McCarthy et al., eds., *Climate Change 2001: Impacts, Adaptation, and Vulnerability*, Contribution of Working Group II to the Third Assessment Report of the IPCC (Cambridge, UK: Cambridge University Press, 2001), Chapter 10.

17. Ibid.
18. Ibid., Chapter 11.
19. Population from U.S. Bureau of the Census, *International Data Base*, electronic database, Suitland, MD, updated 10 October 2002; carbon emissions from sources in note 4.
20. Census Bureau, op. cit. note 20; carbon emissions from sources in note 4.
21. Census Bureau, op. cit. note 20; carbon emissions from sources in note 4.
22. Colum Lynch, "EU Ratifies Global Warming Treaty; Kyoto Accord En Route to Becoming Law Despite U.S. Rejection," *Washington Post*, 1 June 2002; Howard French, "Japan Ratifies Kyoto Pact and Urges U.S. Backing," *New York Times*, 5 June 2002; United Nations Framework Convention on Climate Change (UNFCCC) Secretariat, "Kyoto Protocol Receives 100th Ratification," press release (Bonn, Germany: 18 December 2002).
23. UNFCCC Secretariat, "Kyoto Protocol: Status of Ratification," updated 28 January 2003, at <unfccc.int>, viewed 10 February 2003.
24. UNFCCC Secretariat, op. cit. note 22.

ECONOMIC GROWTH INCHES UP (pages 44–45)

1. Angus Maddison, *The World Economy: A Millennial Perspective* (Paris: Organisation for Economic Co-operation and Development (OECD), 2001), pp. 272–321, with updates from International Monetary Fund (IMF), *World Economic Outlook Database* (Washington, DC: December 2002).
2. Ibid.
3. Percentage of gross world product (GWP) and increase from IMF, op. cit. note 1; drivers from idem, *World Economic Outlook 2002* (Washington, DC: 2002), pp. 3, 21.
4. GWP from IMF, op. cit. note 1; drivers from IMF, op. cit. note 3, p. 29.
5. GWP from IMF, op. cit. note 1; drivers from IMF, op. cit. note 3, pp. 9, 38.
6. GWP from IMF, op. cit. note 1; population from U.S. Bureau of the Census, *International Data Base*, electronic database, Suitland, MD, updated 10 October 2002.
7. Census Bureau, op. cit. note 6; GWP from IMF, op. cit. note 1.
8. Jeffrey Sachs, "Rapid Population Growth Saps Development," *Science*, 19 July 2002, p. 341.
9. OECD, *The Well-being of Nations: The Role of Human and Social Capital* (Paris: 2001), pp. 10–11.
10. Ibid.
11. Ibid.
12. Clifford Cobb, Mark Glickman, and Craig Cheslog, "The Genuine Progress Indicator 2000 Update," Issue Brief (Oakland, CA: Redefining Progress, December 2001); Mathis Wackernagel, Redefining Progress, e-mail to author, 20 January 2003.
13. Robert Costanza et al., "The Value of the World's Ecosystem Services and Natural Capital," *Nature*, 15 May 1997, pp. 253–54.
14. Partha Dasgupta, "Is Contemporary Economic Development Sustainable?" *Ambio*, June 2002, pp. 269–71.
15. Mathis Wackernagel et al., "Tracking the Ecological Overshoot of the Human Economy," *Proceedings of the National Academy of Sciences*, 9 July 2002, p. 9268. The productivity of "global hectares" is an average based on the productivity of ecosystems used by humans.
16. Wackernagel et al., op. cit. note 15.
17. Ibid.; World Wildlife Fund, U.N. Environment Programme, and Redefining Progress, *Living Planet Report 2002*, at <www.panda.org/news_facts/publications/general/livingplanet/index.cfm>, viewed 3 January 2003, p. 21.
18. Peter M. Vitousek et al., "Human Domination of Earth's Ecosystems," *Science*, 25 July 1997, pp. 494–95.
19. Mathis Wackernagel, Chad Monfreda, and Diana Deumling, "Ecological Footprint of Nations: November 2002 Update," Sustainability Issue Brief (Oakland, CA: Redefining Progress, November 2002). Although the average land available to each person globally is 1.9 hectares, the United States is less densely populated, so it has access to more resources, thus allowing 5.3 hectares of resources per person.
20. Calculation based on Wackernagel et al., op. cit. note 19.

FOREIGN DEBT DECLINES (pages 46–47)

1. World Bank, *Global Development Finance 2002* (Washington, DC: 2002), pp. 188–89.
2. Ibid.
3. Ibid., pp. 188–217.
4. Ibid., p. 34.
5. Ibid., p. 36.
6. Anne O. Krueger, *A New Approach to Sovereign Debt*

Restructuring (Washington, DC: International Monetary Fund (IMF), 2002).

7. Chris Kraul, "Costly Lessons of Argentina's Financial Folly," *Los Angeles Times*, 13 January 2002; Daniel Altman, "As Global Lenders Refocus, a Needy World Waits," *New York Times*, 17 March 2002.
8. Paul Blustein, "IMF Refines Bankruptcy Plan After Creditors Object," *Washington Post*, 9 January 2003.
9. World Bank, op. cit. note 1, pp. 190–201.
10. Jeffrey Sachs et al., *Implementing Debt Relief for the HIPCs* (Cambridge, MA: Harvard University Center for International Development, August 1999).
11. Ibid., p. 4.
12. World Bank, *World Debt Tables* 1992–93 (Washington, DC: 1992), pp. 56–59.
13. IMF and World Bank, "Debt Relief Under the Heavily Indebted Poor Countries (HIPC) Initiative," factsheet, August 2002, at <www.imf.org/external/np/exr/facts/hipc.htm>, viewed 21 January 2003; Jubilee campaign from U.N. Development Programme, *Human Development Report 2002* (New York: 2002), pp. 102–04.
14. IMF and World Bank, *Debt Relief for Poverty Reduction: The Role of the Enhanced HIPC Initiative* (Washington, DC: August 2001).
15. World Bank, "HIPC Initiative: Status of Country Cases Considered Under the Initiative," at <www.worldbank.org/hipc/progress-to-date/status_table_Jan03.pdf>, viewed 30 January 2003.
16. Daniel Cohen, *The HIPC Initiative: True and False Promises*, Development Centre Technical Paper No. 166 (Paris: Organisation for Economic Co-operation and Development (OECD), September 2000).
17. Ibid.; Nancy Birdsall and John Williamson, with Brian Deese, *Delivering on Debt Relief: From IMF Gold to a New Aid Architecture* (Washington, DC: Center for Global Development and Institute for International Economics, 2002); David Malin Roodman, *Still Waiting for the Jubilee: Pragmatic Solutions for the Third World Debt Crisis*, Worldwatch Paper 155 (Washington, DC: Worldwatch Institute, April 2001); International Financial Institution Advisory Commission (The Meltzer Commission), *Final Report to the U.S. Congress and Department of Treasury*, 8 March 2000, at <www.bicusa.org/usgovtoversight/meltzer.htm>.
18. Corrupt governments from William Easterly, *The Elusive Quest for Growth: Economists' Adventures and Misadventures in the Tropics* (Cambridge: The MIT Press, 2001), pp. 241–52; channeling debt relief to combat AIDS from Sachs et al., op. cit. note 10.
19. Kunibert Raffer and H. W. Singer, *The Economic North-South Divide: Six Decades of Unequal Development* (Northampton, MA: Edward Elgar, 2001), pp. 197–235; OECD, *Agricultural Policies in OECD Countries 2002* (Paris: 2002), p. 22.
20. World Bank, *Global Economic Prospects and the Developing Countries 2003* (Washington, DC: 2003), p. 10.
21. UNDP, op. cit. note 13, pp. 118–21.

ADVERTISING SPENDING STAYS NEARLY FLAT (pages 48–49)

1. Bob Coen, *Universal McCann's Insider's Report on Advertising Expenditures*, December 2000, at <www.mccann.com/insight/bobcoen.html>, viewed 13 January 2003; table from Bob Coen, *Estimated World Advertising Expenditures*, at <www.mccann.com/insight/bobcoen.html>, viewed 13 January 2003; Coen's estimates include major media (television, radio, press, cinema, Internet, and outdoor ads), as well as direct mail, Yellow Pages, and other forms of media; major media estimate from Zenith Optimedia, *Advertising Expenditures Forecast*, press release, 9 December 2002, at <www.zenithmedia.com/pubopt00.htm>, viewed 14 January 2003.
2. Coen, *Insider's Report*, op. cit. note 1; Zenith Optimedia, op. cit. note 1.
3. Growth from Coen, *Insider's Report*, op. cit. note 1; congressional elections from Zenith Optimedia, op. cit. note 1.
4. Coen, *Insider's Report*, op. cit. note 1.
5. Adam Smith, Zenith Optimedia, e-mail to author, 15 January 2003.
6. Ibid.
7. Ibid.
8. Coen, *Insider's Report*, op. cit. note 1; population from U.S. Bureau of the Census, *International Data Base*, electronic database, Suitland, MD, updated 10 October 2002.
9. Smith, op. cit. note 5; Census Bureau, op. cit. note 8.
10. Laurel Wentz, "Global Marketers Spend $71 Billion," *Advertising Age*, 11 November 2002.
11. Production from DRI Automotive Group, Global Insight, *Global Production of Light Vehicles by Region & Country December 2002* (London: 2002), received from Colin Couchman, e-mail to Michael Renner, 20 January 2003; global fleet from Colin Couchman, DRI Automotive Group, Global Insight, Lon-

don, e-mail to Michael Renner, 21 January 2003.

12. G. Tyler Miller, Jr., *Living in the Environment: Principles, Connections, and Solutions* (Pacific Grove, CA: Brooks/Cole Publishing Company, 1999), p. 732.
13. Meredith Rosenthal et al., "Promotion of Prescription Drugs to Consumers," *New England Journal of Medicine*, 14 February 2002, p. 500.
14. Suz Redfearn, "Journal: Drug Sales Based on 'Seriously Biased' Data," *Washington Post*, 4 June 2002.
15. Rich Thomaselli, "47% of Doctors Feel Pressured by DTC Drug Advertising," *Advertising Age*, 14 January 2003.
16. Euromonitor International, *Marketing to Children: A World Survey*, 2001 Edition (London: 2001), received from Katya Fay, Euromonitor, e-mail to author, 17 January 2003.
17. Thomas N. Robinson et al., "Effects of Reducing Television Viewing on Children's Requests for Toys: A Randomized Controlled Trial," *Journal of Developmental and Behavioral Pediatrics*, June 2001, p. 179.
18. Types of ads from Victor Strasburger, "Children and TV Advertising, Nowhere to Run, Nowhere to Hide," *Journal of Developmental and Behavioral Pediatrics*, June 2001, p. 185; Dina Borzekowski and Thomas Robinson, "The 30-second Effect: An Experiment Revealing the Impact of Television Commercials on Food Preferences of Preschoolers," *Journal of the American Dietetic Association*, January 2001.
19. James D. Sargent et al., "Effect of Seeing Tobacco Use in Films on Trying Smoking Among Adolescents: Cross Sectional Study," *British Medical Journal*, 15 December 2001, p. 1396.
20. Strasburger, op. cit. note 17, pp. 185–86.
21. Ingrid Jacobsson, "Advertising Ban and Children: Children Have the Right to Safe Zones," *Current Sweden*, June 2002.
22. Ibid.
23. Alina Tugend, "Cigarette Makers Take Anti-Smoking Ads Personally," *New York Times*, 27 October 2002.

TOURISM GROWING BUT STILL SHAKY (pages 50–51)

1. Rosa Songel, Statistics and Economic Measurement of Tourism, World Tourism Organization (WTO), e-mails to author, 25 November 1999 and 27 January 2000; WTO, "World Tourism in 2002: Better Than Expected," press release (Madrid: 27 January 2003).
2. WTO, "Tourism Proves As a Resilient and Stable Economic Sector," press release (Madrid: 18 June 2002). The only other decline was in 1982, during the time of the second oil crisis, conflicts in the Falkland Islands and between Lebanon and Israel, and the advent of martial law in Poland.
3. WTO, op. cit. note 2.
4. Ibid.; Songel, op. cit. note 1. Receipts exclude earnings from international transport.
5. World Travel and Tourism Council (WTTC), *The Travel & Tourism Economy 2002, Special End of Year Update* (London: 2002), p. 1.
6. International Civil Aviation Organization, "One Year After 11 September Events ICAO Forecasts World Air Passenger Traffic Will Exceed 2000 Levels In 2003," press release (Montreal: 10 September 2002).
7. International Air Transport Association, "World Air Transport Statistics 2002 Is Out Now," press release (Montreal: 15 July 2002).
8. US Airways, "US Airways to Complete Restructuring Plan in Chapter 11 Reorganization," press release (Arlington, VA: 11 August 2002); United Airlines, "UAL Corp. Files for Chapter 11 Reorganization," press release (Chicago: 9 December 2003).
9. Giovanni Bisignani, "The Aviation Industry Has Proven Its Resilience in 2002," at <www.iata.org/index.htm>, viewed 9 January 2003.
10. WTO, op. cit. note 1.
11. Ibid.; Songel, op. cit. note 1.
12. WTO, op. cit. note 1; Songel, op. cit. note 1.
13. WTO, op. cit. note 1.
14. Ibid.; Songel, op. cit. note 1.
15. WTO, "Leaner, More Competitive Tourism Sector Emerging: East Asia/Pacific Arrivals Set to Double in Next Ten Years," press release (Madrid: 9 January 1999).
16. WTO, *Tourism Highlights 1999* (Madrid: 1999); WTO, op. cit. note 1.
17. WTO, op. cit. note 2.
18. WTO, op. cit. note 1.
19. WTTC, op. cit. note 5.
20. Ibid.
21. Leakage is a WTO estimate cited in David Diaz Benavides, UNCTAD, *The Sustainability of International Tourism in Developing Countries*, paper presented at the Seminar on Tourism Policy and Economic Growth, Berlin, 6–7 March 2001, pp. 8–9.
22. Lisa Mastny, *Traveling Light: New Paths for International Tourism*, Worldwatch Paper 159 (Washington, DC: Worldwatch Institute, December 2001).
23. Ibid.

24. Ibid.
25. International Hotels Environment Initiative, "Six Good Reasons for Going Green," at <www.ihei.org>, viewed 24 January 2000; U.N. Environment Programme, "UNEP Hosts Meeting to Develop Tour Operator Initiative for Sustainable Tourism Development," press release (Nairobi: 6 July 1999).
26. Rainforest Alliance, "Sustainable Tourism Stewardship Council, at <www.rainforest-alliance.org/programs/sv/stsc.html>, viewed 9 January 2002.

WORLD HERITAGE SITES RISING STEADILY (pages 52–53)

1. Cumulative numbers are as of July 2002 and reflect the sum total of the number of sites inscribed each year; United Nations Educational, Scientific, and Cultural Organization (UNESCO), "The World Heritage List," at <whc.unesco.org/nwhc/pages/sites/s_f2.htm>, viewed 11 December 2002.
2. UNESCO, *Properties Inscribed on the World Heritage List* (Paris: July 2002).
3. UNESCO, "Mission Statement," at <whc.unesco.org/nwhc/pages/doc/main.htm>, viewed 15 November 2002.
4. UNESCO, op. cit. note 1; idem, op. cit. note 3.
5. UNESCO, "World Heritage Committee Inscribes 9 New Sites on the World Heritage List," press release (Paris: 27 June 2002).
6. UNESCO, op. cit. note 1.
7. Ibid.
8. UNESCO, "A Brief History," at <whc.unesco.org/nwhc/pages/doc/main.htm>, viewed 15 November 2002.
9. Ibid.
10. UNESCO, "The World Heritage Convention: 30 Years Old and Going Strong," at <whc.unesco.org/venice2002/edito.htm>, viewed 16 December 2002.
11. Ibid.
12. Ibid.; UNESCO, "Heritage: A Gift from the Past to the Future," at <whc.unesco.org/nwhc/pages/doc/main.htm>, viewed 15 November 2002.
13. UNESCO, op. cit. note 10.
14. Ibid.
15. UNESCO, "The List of World Heritage in Danger," at <whc.unesco.org/nwhc/pages/doc/main.htm>, viewed 15 November 2002.
16. Ibid.
17. Ibid.
18. Ibid.
19. UNESCO, op. cit. note 10.
20. UNESCO, op. cit. note 3.
21. Conservation International, "UN Foundation & Conservation International Forge $15 Million Partnership to Protect Global Biodiversity," press release (Washington, DC: 15 November 2002).

VEHICLE PRODUCTION INCHES UP (pages 56–57)

1. DRI Automotive Group, Global Insight, *Global Production of Light Vehicles by Region & Country December 2002* (London: 2002), received from Colin Couchman, e-mail to author, 20 January 2003.
2. Ibid.
3. Ibid.; DRI-WEFA, Global Automotive Group, *Global Sales of Light Vehicles by Region & Country December 2001* (London: 2001); Standard and Poor's DRI, *World Car Industry Forecast Report*, December 2000 and December 1999 (London: 2000 and 1999); American Automobile Manufacturers Association, *World Motor Vehicle Facts and Figures 1998* (Washington, DC: 1998).
4. DRI Automotive Group, op. cit. note 1.
5. Colin Couchman, DRI Automotive Group, Global Insight, London, e-mail to author, 21 January 2003.
6. Calculated from Ward's Communications, *Ward's Motor Vehicle Facts & Figures 2002* (Southfield, MI: 2002).
7. Worldwatch calculation, based on DRI Automotive Group, op. cit. note 1.
8. Ibid.
9. Jason Mark, *Automaker Rankings: The Environmental Performance of Car Companies* (Cambridge, MA: Union of Concerned Scientists, September 2002), p. 1.
10. John DeCicco and Feng An, *Automakers' Corporate Carbon Burden* (Washington, DC: Environmental Defense Fund, July 2002), Figure 9. These percentages are based on conditions in the United States.
11. Oak Ridge National Laboratory (ORNL), *Transportation Energy Databook 22* (Oak Ridge, TN: October 2002), Tables 7.18 and 7.19.
12. Robert Bamberger, "Automobile and Light Truck Fuel Economy: Is CAFE Up to Standards?" Congressional Research Service, Issue Brief for Congress, Washington, DC, 26 July 2002; David E. Rosenbaum, "Senate Deletes Higher Mileage Standard in Energy Bill," *New York Times*, 14 March 2002; Danny Hakim, "Tougher Rules Are Proposed for Gas Mileage," *New York Times*, 13 December 2002.
13. Danny Hakim, "Bush Proposal May Cut Tax on

S.U.V.'s for Business," *New York Times*, 21 January 2003.

14. DeCicco and An, op. cit. note 10, Table A-1.
15. Ibid.
16. ORNL, op. cit. note 11, Figure 6.1.
17. Calculated from Federal Highway Administration, *Highway Statistics 2000* (Washington, DC: 2000). Statistics are for the late 1990s.
18. Divergent fuel efficiency from Michael Renner, "Vehicle Production Declines Slightly," in Worldwatch Institute, *Vital Signs 2002* (New York: W.W. Norton & Company, 2002), pp. 74–75.
19. U.S. Energy Information Administration, *International Energy Outlook 2002* (Washington, DC: U.S. Department of Energy, March 2002), Table E2.
20. Ibid.
21. DeCicco and An, op. cit. note 10, Figure 12.
22. Ibid., Table 3.
23. Mark, op. cit. note 9, p. 10.
24. Danny Hakim, "S.U.V. From Toyota in 2004 To Use Hybrid Technology," *New York Times*, 2 January 2003.
25. Therese Langer and Daniel Williams, *Greener Fleets: Fuel Economy Progress and Prospects* (Washington, DC: American Council for an Energy-Efficient Economy, December 2002), p. 4.

BICYCLE PRODUCTION SEESAWS (pages 58–59)

1. Global production from "World Market," in *Bicycle Retailer and Industry News*, 1 January 2003, p. 23, and from United Nations, *Industrial Commodity Statistics Yearbook 2000* (New York: 2000).
2. John Crenshaw, "Shift to China Makes Waves in Global Market," *Bicycle Retailer and Industry News*, 1 January 2003, p. 22.
3. Megan Hjermstad, "After a Decade of Change, Suppliers Still Restless," *Bicycle Retailer and Industry News*, 1 January 2003, p. 18.
4. Ibid.
5. "Bike Suppliers Work Off Industry Glut," *Bicycle Retailer and Industry News*, 15 March 2002.
6. Crenshaw, op. cit. note 2, p. 22.
7. Based on data in "World Market," op. cit. note 1, and on United Nations, op. cit. note 1.
8. Crenshaw, op. cit. note 2.
9. Use calculated from data in "World Market," op. cit. note 1.
10. Crenshaw, op. cit. note 2.
11. Hjermstad, op. cit. note 3.
12. Ibid.
13. Ghazal Badiozamani, Project Coordinator, UN Car-Free Days, U.N. Department of Economic and Social Affairs, discussion with author, 25 February 2003. Car-free days range from the closing of a few streets on Sundays to the closing of major arteries on weekdays or the more comprehensive banning of private automobiles throughout most of the city, as practiced in Bogotá.
14. See <www.un.org/News/Press/docs/2002/envdev618.doc.htm>.
15. "Bicycle Use," in *Sustainable Transport e-Update*, at <www.itdp.org/STe/STe2/index.html>, viewed 22 January 2003.
16. John Pucher and Lewis Dijkstra, "Making Walking and Cycling Safer: Lessons from Europe," *Transportation Quarterly*, summer 2000, pp. 25–50.
17. Ibid.
18. Ibid.

COMMUNICATIONS NETWORKS EXPAND (pages 60–61)

1. Figures for 1985–90 from International Telecommunication Union (ITU), *World Telecommunication Indicators '98*, Socioeconomic Time-series Access and Retrieval System (STARS) database, downloaded 24 August 1999; 1991–2002 from idem, "Key Global Telecom Indicators for the World Telecommunication Service Sector," Geneva, at <www.itu.int/ITU-D/ict/statistics/at_glance/KeyTelecom99.html>, viewed 4 February 2003.
2. ITU, *World Telecommunication Indicators '98*, op. cit. note 1; idem, "Key Global Telecom Indicators," op. cit. note 1.
3. Host computer count from Internet Software Consortium (ISC), "Internet Domain Survey: Number of Internet Hosts," at <www.isc.org/ds/host-count-history.html>, viewed 4 February 2003. A single host can wire several computers to the Internet. Number of users from Nua, Ltd., "How Many Online?" at <www.nua.ie/surveys/how_many_online/index.html>, viewed 4 February 2003. Users are individuals who use the Internet on a weekly basis; as the user estimates can vary, host computers provide a more reliable measure of the Internet's reach.
4. Population in 1992 and 2002 from U.S. Bureau of the Census, *International Data Base*, electronic database, Suitland, MD, updated 10 October 2002; mobile phone and Internet users in 1992 from ITU,

World Telecommunication Development Report (Geneva: 2002), pp. 6–7, and in 2002 from ITU, "Key Global Telecom Indicators." op. cit. note 1, and from ISC, op. cit. note 3.
5. ITU, op. cit. note 4, pp. 7, 13.
6. Frances Cairncross, *The Death of Distance: How the Communications Revolution Will Change Our Lives* (Cambridge, MA: Harvard Business School Publishing, 1997).
7. ITU, *Internet for a Mobile Generation* (Geneva: 2002).
8. John Markoff, "More Cities Set Up Wireless Networks," *New York Times*, 6 January 2003.
9. "The Fight for Digital Dominance," *The Economist*, 23 November 2002.
10. H. Asher Bolande, "Handsets from China Driving Down Prices," *Wall Street Journal*, 30 January 2003.
11. ITU, op. cit. note 4, p. 17; Tim Kelly, ITU, e-mail to author, 14 February 2003.
12. ITU, op. cit. note 4, pp. 56–57.
13. Ibid.
14. Ibid., p. 15.
15. Ibid., pp. 15–16.
16. ITU, "Cellular Subscribers" (Geneva: 12 December 2002).
17. ITU, op. cit. note 4, p. 27.
18. Joanna Slater, "Computing For All," *Far Eastern Economic Review*, 24 October 2002.
19. Cait Murphy, "The Hunt for Globalization That Works," *Fortune*, 28 October 2002.

SEMICONDUCTOR SALES REBOUND SLIGHTLY (pages 62–63)

1. Semiconductor Industry Association (SIA), "Worldwide Semiconductor Shipments," at <www.semichips.org/pre_statistics.cfm>, viewed 13 August 2002.
2. Ibid.
3. Ibid.
4. Ibid.; World Bank, *World Development Indicators 2001* (Washington, DC: 2001), p. 210.
5. SIA, op. cit. note 1.
6. E. S. Browning and Ianthe Jeanne Dugan, "Stocks Unwound: Aftermath of a Market Mania," *Wall Street Journal*, 16 December 2002.
7. Jeff Chappell, "A Withering Year; Bust 2001, Boom 2000 Just Part of Industry Cycle," *Electronic News*, 8 October 2001, Special Issues section, p. 16.
8. SIA, op. cit. note 1.
9. Ibid.
10. Fabless Semiconductor Association, "Wafer and Packaging Demand Survey," at <www.fsasurvey.com/2002/a/about/execsummary.asp>, viewed 16 December 2002.
11. World Semiconductor Trade Statistics, "WSTS Semiconductor Market Forecast World," at <www.wsts.org/press.html>, viewed 10 November 2002.
12. SIA, "Annual Survey of Work Injuries and Illnesses of the Occupational Health System of the SIA," August 1999, at <www.ohsys.com/OHS%201999%20Survey%20-%20International%20Data.PDF>, viewed 9 December 2002.
13. Jodi Shelton, "Drivers to the Fabless Semiconductor Industry: Analysis of the Fabless Semiconductor Industry," *Future Fab*, 8 July 2002.
14. Ibid.
15. Ann Hwang, "Semiconductors Have Hidden Costs," in Worldwatch Institute, *Vital Signs 2002* (New York: W.W. Norton & Company, 2002), pp. 110–11.
16. Eric D. Williams, Robert U. Ayres, and Miriam Heller, "The 1.7 Kilogram Microchip: Energy and Material Use in the Production of Semiconductor Devices," *Environmental Science and Technology*, 15 December 2002, pp. 5504–09.
17. Ibid.
18. Ibid.
19. Basel Action Network and Silicon Valley Toxics Coalition, *Exporting Harm: The High-Tech Trashing of Asia* (Seattle, WA, and San Jose, CA: February 2002), p. 7.
20. H. Scott Matthews et al., *Disposition and End-of-Life Options for Personal Computers* (Pittsburgh, PA: Carnegie Mellon University Green Design Initiative, July 1997); Basel Action Network and Silicon Valley Toxics Coalition, op. cit. note 19, p. 4.

POPULATION GROWTH SLOWS (pages 66–67)

1. U.S. Bureau of the Census, *International Data Base*, electronic database, Suitland, MD, updated 10 October 2002.
2. Ibid.
3. Ibid.
4. United Nations, *World Population Prospects: The 2002 Revision, Highlights* (New York: 2003), pp. 1–9.
5. "The Health Divide in Europe," *Population Today*, November/December 2002, p. 5; World Health Organization, "*The European Health Report 2002* (Copenhagen: WHO Regional Office for Europe,

2002), pp. 11–15.

6. United Nations, *World Population Prospects: The 2000 Revision, Highlights* (New York: 2001), p. 7.
7. Ibid.
8. Ibid., p. 22.
9. U.N. Population Fund (UNFPA), *State of the World Population 2002* (New York: 2002), pp. 20–24.
10. Ibid.
11. East-West Center, *The Future of Population Growth in Asia* (Honolulu: 2002), p. 104.
12. UNFPA, op. cit. note 9, p. 20.
13. East-West Center, op. cit. note 11, pp. 97–109.
14. United Nations, *World Urbanization Prospects: The 2001 Revision—Data Tables and Highlights* (New York: 2002).
15. Ibid.
16. Ibid.
17. United Nations, op. cit. note 4.
18. Ibid., p. 70.
19. Peter G. Peterson, "The Shape of Things to Come: Global Aging in the Twenty-First Century," *Journal of International Affairs*, fall 2002; David R. Francis, "Europeans Struggle With Idea of 'Replacement Migration,'" *Christian Science Monitor*, 23 May 2002.

HIV/AIDS PANDEMIC SPREADS FURTHER (pages 68–69)

1. Joint United Nations Programme on HIV/AIDS (UNAIDS), *AIDS Epidemic Update: December 2002* (Geneva: 2002), p. 3.
2. Ibid. For the Table, cumulative infections and deaths are calculated by adding annual infections and deaths to data for the previous year. Cumulative infections minus cumulative deaths for 2002 does not equal 42 million (the 2002 estimate of people living with HIV/AIDS) because the dataset does not contain all the revisions made by UNAIDS of data for past years.
3. UNAIDS, op. cit. note 1.
4. World Health Organization, "Women and HIV and Mother-to-Child Transmission," fact sheet, at <www.who.int/health-services-delivery/hiv_aids/English/fact-sheet-10/index.html>, viewed 17 December 2002.
5. Ibid.
6. UNAIDS, *AIDS Epidemic Update: December 2001* (Geneva: 2001), p. 22.
7. UNAIDS, op. cit. note 1, p. 6; Judith Achieng, "HIV/AIDS Remains the Number One Killer," *Inter Press Service*, 23 July 1999.
8. Karen A. Stanecki, *The AIDS Pandemic in the 21st Century*, draft report, XIV International Conference on AIDS, Barcelona (Washington, DC: U.S. Agency for International Development and U.S. Bureau of the Census, July 2002), p. 4.
9. James T. Morris, "Crisis in Africa: The Political Dimensions of Hunger," Statement to the U.N. Security Council, New York, 3 December 2002.
10. Oxfam International and Save the Children UK, *HIV/AIDS and Food Insecurity in Southern Africa* (Oxford and London: 2002), pp. 2–4.
11. UNAIDS, op. cit. note 1, p. 12.
12. Ibid., pp. 14–15.
13. Ibid., p. 14.
14. Ibid., p. 7.
15. Ibid., p. 8; UNAIDS, *Report on the Global HIV/AIDS Epidemic* (Geneva: July 2002), p. 29.
16. UNAIDS, op. cit. note 1, pp. 7–8.
17. Ibid., pp. 4, 29; UNAIDS, op. cit. note 15, pp. 22, 28–29, 32, 35, 37, and 39.
18. UNAIDS, "Accelerating Access to Treatment and Care," fact sheet, at <www.unaids.org/barcelona/presskit/factsheets/FSaccess_en.htm>, viewed 17 December 2002.
19. UNAIDS, op. cit. note 1, p. 18.
20. Ibid., p. 21.
21. UNAIDS, "Meeting the Need," fact sheet, at <www.unaids.org/worldaidsday/2002/press/factsheets/FSneed_en.doc>, viewed 17 December 2002.
22. Ibid.

CIGARETTE PRODUCTION DIPS SLIGHTLY (pages 70–71)

1. U.S. Department of Agriculture (USDA), *Production, Supply, and Distribution*, electronic database, updated 27 November 2002.
2. Ibid.; population data from U.S. Bureau of the Census, *International Data Base*, electronic database, Suitland, MD, updated 10 October 2002.
3. USDA, op. cit. note 1.
4. Ibid.
5. Ibid. Consumption of cigarettes is a residual number based on total production plus imports minus exports. Thus this number includes stockpiled cigarettes.
6. Ibid.
7. USDA, Foreign Agricultural Service, *Russian Federation Tobacco and Products Annual 2002*, at <www.fas

.usda.gov/gainfiles/200205/145683453.pdf>, viewed 24 January 2003; Census Bureau, op. cit. note 2.

8. C. K. Gajalakshmi et al., "Global Patterns of Smoking and Smoking-Attributable Mortality," in Prabhat Jha and Frank Chaloupka, eds., *Tobacco Control in Developing Countries* (Oxford: Oxford University Press, 2000), p. 16.
9. Population from Judith Mackay and Michael Eriksen, *The Tobacco Atlas* (Geneva: World Health Organization (WHO), 2002), p. 34; aggressive marketing from Pan American Health Organization, *Profits Over People: Tobacco Industry Activities to Market Cigarettes and Undermine Public Health in Latin America and the Caribbean*, at <www.paho.org/English/HPP/HPM/TOH/profits_over_people.htm>, viewed 29 December 2002.
10. Death toll from WHO, *The World Health Report, 2002* (Geneva: 2002), p. 65; 1 in 10 from Prabhat Jha and Frank Chaloupka, *Curbing the Epidemic: Governments and the Economics of Tobacco Control* (Washington: World Bank, 1999), p. 1; illnesses from Mackay and Eriksen, op. cit. note 9, p. 32.
11. Deaths in 2030 from Jha and Chaloupka, op. cit. note 10, p. 1; 7 in 10 from Gajalakshmi et al., op. cit. note 8, p. 35.
12. Bruce N. Leistikow et al., "Fire Injuries, Disasters, and Costs from Cigarettes and Cigarette Lights: A Global Overview," *Preventive Medicine*, August 2000, pp. 95–96.
13. "Annual Smoking-Attributable Mortality, Years of Potential Life Lost, and Economic Costs—United States, 1995–1999," *Morbidity and Mortality Weekly Report*, 12 April 2002, p. 303.
14. Mackay and Eriksen, op. cit. note 9, p. 34.
15. "Chair's Text of a Framework Convention on Tobacco Control," WHO, 13 January 2003, at <www.who.int/gb/fctc/PDF/inb6 /einb62.pdf>, viewed 19 January 2003.
16. Clive Bates, "International Tobacco Treaty: Public Health Advocates Face an Uphill Battle," *CorpWatch*, 15 October 2002.
17. Kenneth E. Warner, "Tobacco," *Foreign Policy*, May/June 2002, p.22.
18. Frank J. Chaloupka et al., "The Taxation of Tobacco Products," in Jha and Chaloupka, op. cit. note 8, pp. 242–44.
19. Henry Saffer, "Tobacco Advertising and Promotion," in Jha and Chaloupka, op. cit. note 8, p. 233.
20. Debra Martens, "Graphic Tobacco Warnings Having Desired Effects," *Canadian Medical Association Journal*, 28 May 2002, p. 1453.
21. Federal Trade Commission, *2002 Report on Cigarette Sales, Advertising and Promotion*, at <www.ftc.gov/bcp /menu-tobac.htm>, viewed 14 January 2003.
22. Saffer, op. cit. note 18, p. 215.
23. Ibid., p. 224; 6.3 percent based on analysis of 22 high-income countries for 1970–92.
24. Ibid., p. 231.
25. Seth Mydans, "Thais Impose Wide Ban on Smoking, and, Surprise, It Works," *New York Times*, 19 December 2002.
26. Trevor Woollery et al., "Clean Indoor-Air Laws and Youth Access Restrictions," in Jha and Chaloupka, op. cit. note 8, p. 273.
27. Caroline M. Fichtenberg and Stanton A. Glantz, "Effect of Smoke-free Workplaces on Smoking Behaviour: Systematic Review," *British Medical Journal*, 27 July 2002.
28. Working Group on Tobacco Control of the Federal Provincial Territorial Advisory Committee on Population Health, *The National Strategy: Moving Forward—The 2002 Progress Report on Tobacco Control* (Ottawa: Communication Canada, 2002).

VIOLENT CONFLICTS CONTINUE TO DECLINE (pages 74–75)

1. Arbeitsgemeinschaft Kriegsursachenforschung (AKUF), "Warten auf den Großen Krieg. Weltweit 45 Kriegerische Konflikte im Jahr 2002," press release (Hamburg, Germany: University of Hamburg, 16 December 2002); idem, "Das Kriegsgeschehen 2001 im Überblick," at <www.sozialwiss .uni-hamburg.de/publish/Ipw/Akuf/kriege_aktuell .htm>, viewed 16 December 2002.
2. AKUF, "Warten auf den Großen Krieg," op. cit. note 1; idem, "Das Kriegsgeschehen 2001 im Überblick," op. cit. note 1.
3. Figure 1 from AKUF, "Warten auf den Großen Krieg," op. cit. note 1, from idem, "Das Kriegsgeschehen 2001 im Überblick," op. cit. note 1, and from Wolfgang Schreiber, AKUF, e-mail to author, 19 December 2002.
4. AKUF, "Warten auf den Großen Krieg," op. cit. note 1; idem, "Das Kriegsgeschehen 2001 im Überblick," op. cit. note 1.
5. Anatol Lieven, "The Push for War," *London Review of Books*, 3 October 2002; Jay Bookman, "The Bush Plan for Empire," at <www.bushwatch.net/

empire.htm>, viewed 3 March 2003.

6. Daniel Smith, "The World at War—January 1, 2003," *Defense Monitor*, January/February 2003.
7. Author's estimates based on International Institute for Strategic Studies (IISS), *The Military Balance 2002–2003* (Oxford: Oxford University Press, 2002).
8. Ibid.
9. Coalition to Stop the Use of Child Soldiers, *Child Soldiers Global Report* (London: May 2001); Nicol Degli Innocenti, "About 800,000 Children 'Being Used as Soldiers Worldwide,'" *Financial Times*, 13 June 2001.
10. See Taylor B. Seybolt, "Measuring Violence: An Introduction to Conflict Data Sets," in Stockholm International Peace Research Institute, *SIPRI Yearbook 2002. Armaments, Disarmament and International Security* (New York: Oxford University Press, 2002), pp. 84–85.
11. Nils Petter Gleditsch et al., "Armed Conflict: 1946–2001: A New Dataset," *Journal of Peace Research*, vol. 39, no. 5 (2002), pp. 615–37. In Figure 2, "minor conflicts" involve at least 25 battle-related deaths in a year; if they involve more than 1,000 battle-related deaths during the course of the conflict, they are considered "intermediate conflicts." "Wars" are conflicts with at least 1,000 battle-related deaths in a given year. For most of the "unclear cases," there is uncertainty about whether fatalities during the reporting year surpassed 25 (in addition to other outstanding definitional questions).
12. Heidelberger Institut für Internationale Konfliktforschung (HIIK), *Konfliktbarometer 2002* (Heidelberg, Germany: Institute for Political Science, University of Heidelberg, 2002), p. 3, and earlier editions of this report.
13. Ibid., p. 3.
14. Ibid.
15. Ibid., p. 6; Amy Waldman, "Sri Lanka to Explore a New Government," *New York Times*, 6 December 2002.
16. HIIK, op. cit. note 12, p. 7.
17. PIOOM—Interdisciplinary Research Programme on Causes of Human Rights Violations, *World Conflict & Human Rights Map 2001/2002* (Leiden, Netherlands: 2002).
18. AKUF, "Warten auf den Großen Krieg," op. cit. note 1; idem, "Das Kriegsgeschehen 2001 im Überblick," op. cit. note 1.
19. Cost estimate calculated in 2001 dollars by author on basis of data in "The 2002 Chart of Armed Conflict," wall chart distributed with IISS, op. cit. note 7.

PEACEKEEPING EXPENDITURES DOWN SLIGHTLY (pages 76–77)

1. U.N. Department of Public Information (UNDPI), "United Nations Peacekeeping Operations. Background Note," New York, 15 January 2003; U.N. Department of Peacekeeping Operations (UNDPKO), "UN Peace Operations 2002: Year in Review," at <www.un.org/Depts/dpko/yir/english>, December 2002. (In July 1997, the United Nations switched its peacekeeping accounts from calendar years to July–June reporting periods.) Earlier numbers also based on sources documented in Michael Renner, "Peacekeeping Expenditures Rise Again," in Worldwatch Institute, *Vital Signs 2002* (New York: W.W. Norton & Company, 2002), pp. 96–97.
2. Elisabeth Sköns et al., "Tables of Military Expenditure," in Stockholm International Peace Research Institute (SIPRI), *SIPRI Yearbook 2002. Armaments, Disarmament and International Security* (New York: Oxford University Press, 2002), p. 231.
3. UNDPKO, "Monthly Summary of Contributors," at <www.un.org/Depts/dpko/dpko/contributors/index.htm>, viewed 14 January 2003; personnel number also based on William Durch, Henry Stimson Center, Washington, DC, e-mail to author, 9 January 1996, and on Global Policy Forum, at <www.globalpolicy.org/security/peacekpg/data/pkomctab.htm>, viewed 28 January 2002.
4. UNDPI, op. cit. note 1.
5. UNDPKO, op. cit. note 1.
6. UNDPI, op. cit. note 1.
7. Ibid.
8. Author's calculation, based on data from UNDPKO, op. cit. note 3.
9. UNDPKO, at <www.un.org/Depts/dpko/unamsil/UnamsilF.htm>; UNDPKO, op. cit. note 3.
10. United Nations, "Security Council Expands Authorized Troop Level in Democratic Republic of Congo to 8,700, Noting 'Encouraging Developments' on Ground," press release (New York: 4 December 2002).
11. Rachel L. Swarns, "Congo and Its Rebels Sign Accord to End War," *New York Times*, 18 December 2002.
12. UNDPKO, op. cit. note 3.
13. Ibid.; East Timor independence from "Timeline: East

Timor 1975 to 2002," *BBC News Online*, 17 May 2002.
14. UNDPI, op. cit. note 1.
15. Ibid.
16. "Status of Contributions to the Regular Budget, International Tribunals and Peacekeeping Operations as at 31 December 2002," from Mark Gilpin, Chief of United Nations Contributions Service, New York, letter to author, 23 January 2003.
17. Ibid.; Marjorie Ann Browne, "United Nations Peacekeeping: Issues for Congress," Congressional Research Service, Washington, DC, 19 July 2002.
18. "Status of Contributions," op. cit. note 16.
19. Renata Dwan, Thomas Papworth, and Sharon Wiharta, "Multilateral Peace Missions, 2001," in SIPRI, op. cit. note 2, pp. 130–39; International Institute for Strategic Studies (IISS), "The 2002 Chart of Armed Conflict," wall chart distributed with *The Military Balance 2002–2003* (London: Oxford University Press, 2002).
20. Number of non-UN peacekeeping troops calculated from Dwan, Papworth, and Wiharta, op. cit. note 19, from SIPRI, *SIPRI Yearbook* (New York: Oxford University Press, various years), and from IISS, op. cit. note 19, and previous editions of the wall chart.
21. Compiled from Dwan, Papworth, and Wiharta, op. cit. note 19, from SIPRI, op. cit. note 20, and from IISS, op. cit. note 19, and previous editions of the wall chart.
22. Dwan, Papworth, and Wiharta, op. cit. note 19.
23. Ibid.
24. Ibid.
25. Ibid.
26. United Nations, "Security Council Resolutions," at <www.un.org/Docs/scres/2001/sc2001.htm>, 20 December 2001.
27. IISS, op. cit. note 19.
28. Jim Wurst, "Afghanistan: Security Force Extended One Year; No Change In Mandate," *UN Wire*, 27 November 2002.
29. Calculated from data in IISS, op. cit. note 19.

BIRDS IN DECLINE (pages 82–83)

1. Alison J. Stattersfield and David R. Capper, eds., *Threatened Birds of the World* (Barcelona: Lynx Edicions, 2000), pp. 2–20.
2. Kenneth D. Whitney et al., "Seed Dispersal by *Certatogymna* Hornbills in the Dja Reserve, Cameroon," *Journal of Tropical Ecology*, vol. 14 (1998), pp. 351–71; Josep del Hoyo, Andrew Elliott, and Jordi Sargatal, eds., *Handbook of the Birds of the World, Volume 5* (Barcelona: Lynx Edicions, 1999), pp. 499, 523; Deborah J. Pain and Michael W. Pienkowski, eds., *Farming and Birds in Europe* (London: Academic Press, 1997), pp. 128–37.
3. S. J. Ormerod and Stephanie J. Tyler, "Dippers and Grey Wagtails as Indicators of Stream Acidity in Upland Wales," in A. W. Diamond and F .L. Filion, *The Value of Birds* (Cambridge: International Council for Bird Preservation, 1987), pp. 191–207.
4. John H. Rappole, Scott R. Derrickson, and Zdenek Hubálek, "Migratory Birds and Spread of West Nile Virus in the Western Hemisphere," *Emerging Infectious Diseases* (Atlanta, GA: Centers for Disease Control and Prevention, 2000).
5. Stattersfield and Capper, op. cit. note 1.
6. Ibid.
7. Ibid.
8. Ibid.
9. George W. Cox, *Alien Species in North America and Hawaii: Impacts on Natural Ecosystems* (Washington, DC: Island Press, 1999), pp. 3–35, 281–97.
10. Ibid.
11. David Pimentel et al., "Environmental and Economic Costs Associated with Non-indigenous Species in the United States," *BioScience*, vol. 50, no. 1 (2000), p. 53–65.
12. BirdLife Malta, at <www.birdlifemalta.org>, viewed October 2002.
13. Josep del Hoyo, Andrew Elliott, and Jordi Sargatal, eds., *Handbook of the Birds of the World, Volume 2* (Barcelona: Lynx Edicions, 1994), pp. 336–41; Robert S. Ridgely and Paul J. Greenfield, *The Birds of Ecuador: Status, Distribution, and Taxonomy* (Ithaca, NY: Cornell University Press, 2001), p. 86.
14. Del Hoyo, Elliott, and Sargatal, op. cit. note 13, pp. 533–50.
15. N. Snyder et al., eds., *Parrots: Status Survey and Conservation Action Plan* (Gland, Switzerland, and Cambridge: IUCN–World Conservation Union, 1999).
16. "Longlining: A Major Threat to the World's Seabirds," *World Birdwatch*, June 2000, pp. 10–14; American Bird Conservancy, *Sudden Death on the High Seas* (Washington, DC: 2002).
17. "Agreement on the Conservation of Albatrosses and Petrels," Environment Australia, at <www.ea.gov.au/biodiversity/international/albatross>, viewed December 2002.
18. William Moskoff, "The Impact of Oil Spills on Birds," *Birding*, February 2000, pp. 44–49; Bhushan

Bahree, Carlta Vitzthum, and Erik Portanger, "Tanker Saga Illustrates How Rescues are Hurt by Cross-Current Goals," *Wall Street Journal Europe*, 25 November 2002.

19. Juan Forero, "Ambitions Scaled Back for Ecuador Pipeline," *International Herald Tribune*, 31 October 2002; James V. Grimaldi, "Texas Firms Line Up U.S. Aid in Peru," *Washington Post*, 20 November 2002.
20. John P. McCarty and Anne L. Secord, "Possible Effects of PCB Contamination on Female Plumage Color and Reproductive Success in Hudson River Tree Swallows," *The Auk*, vol. 117, no. 4 (2000), pp. 987–95; idem, "Nest-building Behavior in PCB-Contaminated Tree Swallows," *The Auk*, vol. 116, no. 1 (1999), pp. 55–63.
21. David Pimentel et al., "Environmental and Economic Costs of Pesticide Use," *BioScience*, vol. 42, no. 10 (1992), pp. 750–60.
22. U.S. Fish & Wildlife Service, "Service Continues to Expand Non-toxic Shot Options—Study Shows Ban on Lead Shot Saves Millions of Waterfowl," press release (Washington, DC: 25 October 2000).
23. "U.S.A. Towerkill Summary," at <www.towerkill.com>, viewed March 2002; Guyonne F. E. Janss, "Avian Mortality from Power Lines: A Morphologic Approach of a Species-specific Mortality," *Biological Conservation*, vol. 95, no. 3 (2000), pp. 353–59.
24. Maureen F. Harvey, "Proposed Windfarms Equal Potential Threats to Migrating Songbirds and Raptors," *The Maryland Yellowthroat*, January/February 2003, pp. 5–6.
25. Jeff Price and Patricia Glick, *The Birdwatcher's Guide to Global Warming* (Washington, DC: American Bird Conservancy and National Wildlife Federation, 2002), pp. 12–14.
26. John P. McCarty, "Ecological Consequences of Recent Climate Change," *Conservation Biology*, April 2001, pp. 320–29; Price and Glick, op. cit. note 25.
27. Alison J. Stattersfield et al., *Endemic Bird Areas of the World: Priorities for Biodiversity Conservation* (Cambridge: BirdLife International, 1998), pp. 10–15.
28. Ibid.
29. Alison J. Stattersfield, BirdLife International, e-mail to author, June 2002.
30. Stattersfield et al., op. cit. note 27.
31. C. J. M. Musters et al., "Breeding Birds as a Farm Product," *Conservation Biology*, April 2001, pp. 363–69; M. Ausden and G. J. M. Hirons, "Grassland Nature Reserves for Breeding Wading Birds in England and the Implications for the ESA Agri-environment Scheme," *Biological Conservation*, vol. 106, no. 2 (2002), pp. 279–91; "Prairie Tales: What Happens When Farmers Turn Prairies into Farmland and Farmland into Prairies," *Science News*, 20 January 1996, pp. 44–45.
32. Martha Honey, *Ecotourism and Sustainable Development: Who Owns Paradise?* (Washington, DC: Island Press, 1999), pp. 6–25.
33. Duan Biggs, BirdLife South Africa project coordinator, e-mail to author, August 2002.

SMALL ISLANDS THREATENED BY SEA LEVEL RISE (pages 84–85)

1. For islands with multiple sea level gauges, the data were averaged to obtain a single value. For some islands and chains of islands, however, only one sea level gauge was available and was used to represent the entire island nation. Month-to-month gauge data from the University of Hawaii Sea Level Center, GLOSS database, at <ilikai.soest.hawaii.edu/uhslc/woce.html>, viewed 28 November 2002. Long-term trends derived from the Permanent Service for Mean Sea Level, at <www.nbi.ac.uk/psmsl/datainfo/rlr.trends>, viewed 28 November 2002, and from the South Pacific Sea Level and Climate Monitoring Project, *Pacific Country Reports* (various), at <www.pacificsealevel.org/islandreport.htm>, viewed 28 November 2002.
2. Intergovernmental Panel on Climate Change (IPCC), *Climate Change 2001: The Scientific Basis* (Cambridge: Cambridge University Press, 2001), p. 31.
3. Ibid., p. 4.
4. Jeff Williams and Virginia Burkett, "Forum on Sea-Level Rise and Coastal Disasters," *Soundwaves*, December 2001/January 2002, at <soundwaves.usgs.gov /2002/01/meetings2.html>, viewed 17 November 2002.
5. IPCC, op. cit. note 2, p. 16.
6. IPPC, *Climate Change 2001: Impacts, Adaptation, and Vulnerability* (Cambridge: Cambridge University Press, 2001), p. 847.
7. Ibid., p. 845.
8. Jon Barnett, "Adapting to Climate Change in Pacific Island Countries: The Problem of Uncertainty," *World Development*, vol. 29, no. 6 (2001), p. 978.
9. Ibid.
10. Kalinga Seneviratne, "Tuvalu Steps Up Threat to Sue Australia, U.S." *Pacific Islands Report*, 8 September 2002.
11. John Pernetta, "Rising Seas and Changing Currents,"

People and the Planet, vol. 7, no. 2 (1998); Robert J. Nicholls, "Synthesis of Vulnerability Analysis Studies," *Proceedings of WORLDCOAST '93*, August 1994, p. 34.

12. IPPC, *The Regional Impacts of Climate Change: An Assessment of Vulnerability* (Cambridge: Cambridge University Press, 1998), p. 350.
13. "Small Island States Meet Over Rising Sea Levels," *Environmental News Network*, 14 July 1999.
14. Synthesis and Upscaling of Sea-Level Rise Vulnerability Assessment Studies (SURVAS), *SURVAS Database*, at <www.survas.mdx.ac.uk/content.htm>, viewed 13 November 2002.
15. IPPC, op. cit. note 6, p. 856.
16. IPPC, op. cit. note 12, p. 346.
17. IPPC, op. cit. note 2, p. 4.
18. Barnett, op. cit. note 8, p. 986.
19. IPPC, op. cit. note 6, p. 854.
20. Angie Knox, "Sinking Feeling in Tuvalu," *BBC News Online*, 28 August 2002.
21. Leonard A. Nurse, "Climate Change and Coastal Vulnerability in Small Island States," prepared for the AOSIS Inter-Regional Preparatory Meeting for the World Summit on Sustainable Development, 7–11 January 2002.
22. Ibid.
23. Ibid.
24. IPPC, op. cit. note 6, p. 858.
25. Ibid., p. 862.
26. Ibid.
27. Barbados Programme of Action, *Report of the Global Conference on the Sustainable Development of Small Island Developing States*, October 1994, p. 9.
28. IPCC, op. cit. note 12, p. 336.

RICH-POOR DIVIDE GROWING (pages 88–89)

1. World Bank, *World Development Report 2000/2001* (New York: Oxford University Press, 2001), p. 51.
2. Ibid.
3. U.N. Development Programme, *Human Development Report 1998* (New York: Oxford University Press, 1998), p. 57.
4. Ibid.
5. Giovanni Andrea Cornia and Julius Court, *Inequality, Growth and Poverty in the Era of Liberalization and Globalization* (Helsinki, Finland: UNU World Institute for Development Economics Research, 2001), p. 7.
6. Ibid., pp. 7–8. Gross world product reflects purchasing power parity.
7. Cornia and Court, op. cit. note 5.
8. Ibid., p. 8.
9. Ibid.; Giovanni Andrea Cornia, University of Florence, e-mail to author, 6 February 2003.
10. Cornia and Court, op. cit. note 5, p. 8.
11. World Bank, op. cit. note 1, p. 37.
12. Ibid., p. 65.
13. Ibid., p. 37.
14. Cornia and Court, op. cit. note 5, pp. 9, 12.
15. Ibid., pp. 12, 19.
16. World Bank, *World Development Indicators 2002* (Washington, DC: 2002), pp. 74–76. High-income nations according to World Bank definition.
17. Arthur F. Jones Jr. and Daniel H. Weinberg, *The Changing Shape of the Nation's Income Distribution* (Washington, DC: U.S. Bureau of the Census, 2000), pp. 1–2; U.S. Bureau of the Census, *Measures of Household Income Inequality (Table IE-6)*, at <www.census.gov/hhes/income/histinc/ie6.html>, viewed 7 February 2003.
18. Paul Krugman, "For Richer," *New York Times*, 20 October 2002.
19. World Bank, op. cit. note 1, p. 57.
20. World Bank, op. cit. note 16, p. 124. Low-, middle-, and high-income countries according to World Bank definition.
21. World Bank, cited in U.N. Population Fund, *State of World Population 2002* (New York: 2002), p. 35.
22. World Bank, op. cit. note 16, p. 72.
23. Federal Interagency Forum on Child and Family Statistics, *America's Children: Key National Indicators of Well-Being, 2000* (Washington, DC: U.S. Government Printing Office, 2002), p. 31.
24. Ibid.
25. World Bank, op. cit. note 1, p. 57.
26. Ibid., pp. 57–58.
27. Ibid., p. 55.
28. Ibid., p. 52.
29. Ibid., p. 53.
30. Ibid.
31. Ibid., pp. 55–56.
32. Ibid., pp. 53–54.
33. Ibid., p. 56.
34. Ibid.
35. Cornia and Court, op. cit. 5, p. 24.
36. Ibid., p. 6.
37. World Bank, op. cit. note 16, pp. 74–77.

GAP IN CEO-WORKER PAY WIDENS (pages 90–91)

1. Total compensation, with stock options, from Louis

Lavelle, Frederick F. Jesperson, and Michael Arndt, "Executive Pay," *Business Week Online*, 15 April 2002; 350 times is a Worldwatch calculation based on executive pay from Towers Perrin, "Total Remuneration—Chief Executive Officer," at <www.towers.com/TOWERS/services_products/TowersPerrin/wwtr01/im.../exhibit1.gif>, and on worker pay from "Hourly Direct Pay in U.S. Dollars for Production Workers in Manufacturing, 30 Countries or Areas, 1975–2001," in Bureau of Labor Statistics (BLS), "International Comparisons of Hourly Compensation Costs for Production Workers in Manufacturing, 1975–2001," Supplementary Tables for BLS News Release, 27 September 2002.

2. Fivefold is from Scott Klinger et al., *Executive Excess 2002: CEOs Cook the Books, Skewer the Rest of Us*, Ninth Annual CEO Compensation Survey (Washington, DC: Institute for Policy Studies and United for a Fair Economy, August 2002).
3. Lavelle, Jesperson, and Arndt, op. cit. note 1; "Hourly Direct Pay in U.S. Dollars for Production Workers," op. cit. note 1. Worldwatch conversion of hourly pay to annual pay assumes a work year of 2,000 hours.
4. "Hourly Direct Pay in U.S. Dollars for Production Workers," op. cit. note 1.
5. "Executive Compensation Scoreboard," *Business Week*, 15 April 2002.
6. Lavelle, Jesperson, and Arndt, op. cit. note 1; gap with workers is a Worldwatch calculation based on CEO data in ibid. and on manufacturing worker data from "Hourly Direct Pay in U.S. Dollars for Production Workers," op. cit. note 1.
7. Lavelle, Jesperson, and Arndt, op. cit. note 1.
8. Lucian Arye Bebchuk, Jesse M. Fried, and David I. Walker, "Managerial Power and Rent Extraction in the Design of Executive Compensation," Discussion Paper 366, Harvard John M. Olin Discussion Paper Series (Cambridge, MA: Harvard University, June 2002).
9. Lavelle, Jesperson, and Arndt, op. cit. note 1.
10. Ibid.
11. Eric Wahlgren, "Spreading the Yankee Way of Pay," *Business Week Online*, 18 April 2001.
12. Ibid.
13. Ibid.
14. Klinger et al., op. cit. note 2, p. 7.
15. Lavelle, Jesperson, and Arndt, op. cit. note 1.
16. Klinger et al., op. cit. note 2.
17. Lavelle, Jesperson, and Arndt, op. cit. note 1.
18. Wahlgren, op. cit. note 11.
19. Lavelle, Jesperson, and Arndt, op. cit. note 1.
20. Klinger et al., op. cit. note 2, p. 3.
21. Ibid.
22. Ibid.
23. Lavelle, Jesperson, and Arndt, op. cit. note 1.
24. "Powerful Voices Join the Chorus of Pay Critics," *Charlotte Observer*, 25 August 2002.
25. Representative Martin Slav Sabo, at <www.house.gov/sabo/ie.htm>.

SEVERE WEATHER EVENTS ON THE RISE (pages 92–93)

1. Angelika Wirtz, Munich Reinsurance Company (Munich Re), e-mail to author, 28 January 2003.
2. U.S. National Climatic Data Center (NCDC), National Oceanic and Atmospheric Administration (NOAA), *Climate of 2002 Annual Review*, 23 January 2003, at <www.ncdc.noaa.gov/oa/climate/research/2002/ann/ann02.html>, viewed 1 February 2003.
3. Wirtz, op. cit. note 1.
4. Munich Re, "Press Release 30th December 2002," at <www.munichre.com>, viewed 31 December 2002.
5. Ibid.
6. International Federation of Red Cross and Red Crescent Societies (IFCR), *World Disasters Report 2002* (Geneva: 2002), pp. 195–96; idem, *World Disasters Report 2001* (Geneva: 2001), Chapter 8–Summary.
7. IFCR, *World Disasters Report* 2002, op. cit. note 6; idem, *World Disasters Report 2001*, op. cit. note 6.
8. Munich Re, op. cit. note 4.
9. Munich Re, *Topics: Natural Catastrophes January—September 2002* (draft), e-mail from Nick Nuttall, U.N. Environment Programme (UNEP), to Arunima Dhar, Worldwatch Institute, 8 November 2002.
10. World Meteorological Organization (WMO), "Global Surface Temperatures Second Warmest on Record, Return to El Niño Conditions," press release (Geneva: 17 December 2002).
11. Total losses include direct damage to buildings and other assets and secondary losses due to business interruptions; Munich Re, op. cit. note 4.
12. Moscow deaths from Boris Kagarlitsky, "Ded Moroz's Darker Side," *Moscow Times*, 15 January 2002; "Snowstorm Grips Russia's Far East," *CNN.com*, 8 January 2002.
13. "Floods Cause Havoc in Southern Russia," *BBC News*, 10 January 2002.
14. NCDC, NOAA, "Climate-Watch, February 2002," at

<lwf.ncdc.noaa.gov>, viewed 30 December 2002.

15. U.N. Office for the Coordination of Humanitarian Affairs (UNOCHA), "Peru—Snowstorms OCHA Situation Report No. 3" (Geneva: 23 July 2002).
16. NCDC, op. cit. note 2.
17. Number of deaths from Asian Disaster Reduction Center (ADRC), "ADRC Latest Disaster Information—India: Heatwave: 2002/05/15," (Tokyo: 2002); record death toll from NCDC, NOAA, "Climate-Watch, May 2002," at <lwf.ncdc.noaa.gov>, viewed 30 December 2002.
18. UNOCHA, "India—Floods OCHA Situation Report No. 3" (Geneva: 31 July 2002).
19. "2002 Termed All-India Drought Year," *Business Line*, 5 October 2002.
20. WMO, op. cit. note 10.
21. Comparison to Dust Bowl according to an administrator for the U.S. Department of Agriculture, from Cheryl Runyon, *2002 Drought Report* (Washington, DC: National Conference of State Legislatures, September 2002).
22. Job losses from Australian Bureau of Statistics and cited in NCDC, NOAA, "Climate-Watch, October 2002," at <lwf.ncdc.noaa.gov>, viewed 30 December 2002; loss in economic growth from BBC News and Reuters, cited in NCDC, NOAA, "Climate-Watch, November 2002," at <lwf.ncdc.noaa.gov>, viewed 30 December 2002.
23. UNEP, "Natural Disasters Set to Cost over $70 Billion: Insurers Warn of Mounting Price Tag of Climate Change at COP8 Meeting," press release (Delhi/Nairobi: 29 October 2002).
24. Ibid.
25. UNOCHA, "Kenya—Floods OCHA Situation Report No. 1" (Geneva: 13 May 2002).
26. "Eritrea Issues Drought Warning," *BBC News*, 25 July 2002.
27. World Food Programme, "Millions Threatened with Starvation in Horn of Africa" (Rome: 28 October 2002).
28. NCDC, op. cit. note 2.
29. "Dongting Lake Water Level Recedes," *China Daily*, 26 August 2002.
30. ADRC, "ADRC Latest Disaster Information—China: Flood: 2002/08/22" (Tokyo: 2002).
31. Record rains from WMO, op. cit. note 10; cost from UNEP, op. cit. note 23.
32. "Weather is Getting More Extreme: UN," *Associated Press*, 18 December 2002.
33. Calculated with data from Wirtz, op. cit. note 1.
34. "Global Climate Change Threatens the Insurance Industry," *E/The Environmental Magazine*, August 2002.
35. "Insurers See More Disasters Due to Climate Change," *Reuters*, 2 November 2001.
36. Calculated with data from Wirtz, op. cit. note 1.
37. Calculated with data provided by Angelika Wirtz, Munich Re, e-mail to author, 23 January 2003.
38. IFCR, *World Disasters Report 2001* op. cit. note 6, pp. 83–101.
39. IFCR, *World Disasters Report* 2002, op. cit. note 6, p. 89.
40. Mark Townsend, "Environmental Refugees," *The Ecologist*, June 2002.
41. Töpfer cited in ibid.
42. Christian Brauner, "Climate Research Does Not Remove the Uncertainty: Coping with the Risks of Climate Change," Swiss Re, 1998, p. 7, at <www.swissre.com>, viewed 20 November 2002.
43. J. T. Houghton et al., eds., *Climate Change 2001: The Scientific Basis*, Contribution of Working Group I to the Third Assessment Report of the Intergovernmental Panel on Climate Change (Cambridge, U.K.: Cambridge University Press, 2001).
44. Ibid., p. 13.
45. Brauner, op. cit. note 42, p. 5.
46. Innovest Strategic Value Advisors, *Climate Change & The Financial Services Industry: Module 1 – Threats and Opportunities*, prepared for the UNEP Finance Initiatives Climate Change Working Group (New York: July 2002).
47. UNEP, "Financial Sector, Governments and Business Must Act on Climate Change or Face the Consequences," press release (Zurich/Nairobi: 8 October 2002).
48. Gerhard Berz and Thomas Loster, "Climate Change—Threats and Opportunities for the Financial Sector" (Munich: Munich Re, September 2001).
49. UNEP, op. cit. note 47.

HIGH FARM SUBSIDIES PERSIST (pages 96–97)

1. Organisation for Economic Co-operation and Development (OECD), *Agricultural Policies in OECD, Monitoring and Evaluation 2002* (Paris: 2002), p. 22.
2. Ibid., p. 158.
3. Ibid., pp. 23, 32.
4. Ibid., pp. 23, 160–61.
5. Ibid.
6. Ibid.

7. Ibid., p. 164.
8. OECD, *Distributional Effects of Agricultural Support in Selected OECD Countries* (Paris: 1999).
9. U.S. Department of Agriculture, *Food and Agricultural Policy: Taking Stock for the New Century* (Washington, DC: September 2001), pp. 46–51; OECD, op. cit. note 1, pp. 46–48; idem, op. cit. note 8, pp. 33–36.
10. OECD, op. cit. note 8, p. 18.
11. Environmental Working Group, "About the 2002 Farm Bill: A Missed Opportunity," analysis from farm subsidy database, at <www.ewg.org/farm>.
12. OECD, op. cit. note 1, p. 17.
13. Ibid., p. 17.
14. OECD, *Agricultural Policies in OECD Countries: Monitoring and Evaluation 2001* (Paris: 2001), pp. 25; OECD, op. cit. note 1, pp. 162–63; barriers to diversifying from Thomas L. Dobbs and Jules N. Pretty, *The United Kingdom's Experience with Agri-environmental Stewardship Schemes: Lessons and Issues for the United States and Europe* (Colchester, U.K.: South Dakota State University and University of Essex Centre for Environment and Society, 2001), p. 7.
15. Linda M. Dumke and Thomas L. Dobbs, "Historical Evidence of Crop Systems in Eastern South Dakota: Economic Influences," Economic Research Report 99-2, Department of Economics, South Dakota State University, Brookings, SD, July 1999, pp. 52–53.
16. Norman Myers with Jennifer Kent, *Perverse Subsidies: Tax $s Undercutting Our Economies and Environments Alike* (Winnipeg, MN, Canada: International Institute for Sustainable Development, 1998), pp. 114–17.
17. Ibid., pp. 46–47.
18. OECD, op. cit. note 1, p. 150.
19. IndiaStat website, "Central Subsidy on Food and Fertilisers (1976–77 to 2001–2002)," data provided by IndiaStat.com, e-mail to author, 19 November 2002.
20. Figure of $6.5 billion from OECD, op. cit. note 1, p. 160; Ginger Thompson, "Nafta to Open Foodgates, Engulfing Rural Mexico," *New York Times*, 19 December 2002.
21. Chrispin Inambao, "Cabinet Approves Livestock Subsidies," *The Namibian*, 11 November 2002; OECD, op. cit. note 1, p. 90.
22. OECD, op. cit. note 1, p. 158.
23. Xinshen Diao, Terry Roe, and Agapi Somwaru, "Developing Country Interests in Agricultural Reforms under the World Trade Organization," *American Journal of Agricultural Economics*, August 2002.
24. Oxfam, *Rigged Rules and Double Standards: Trade, Globalisation, and the Fight Against Poverty* (Oxford: 2002), p. 115; Mark Ritchie, Sophia Murphy, and Mary Beth Lake, *United States Dumping on World Agricultural Markets* (Minneapolis, MN: Institute for Agriculture and Trade Policy, 2003).
25. OECD, op. cit. note 1, p. 21.
26. Ibid., p. 22; Elizabeth Olson, "Global Trade Negotiations Are Making Little Progress," *New York Times*, 7 December 2002.
27. Philip Brasher, "House Passes Election-Year Expansion in Farm Subsidies," *Washington Post*, 3 May 2002.
28. Dobbs and Pretty, op. cit. note 14, pp. 1, 17; Elaine Sciolino, "A Fight Over Farms Ends, Opening Way to Wider Europe," *New York Times*, 25 October 2002.
29. Dobbs and Pretty, op. cit. note 14, pp. 1, 17.
30. Sciolino, op. cit. note 28.
31. Margaret Beckett, Patricia Hewitt, and Clare Short, "Money Dished Out to Our Farmers Takes Food from the Mouths of the Poor," (London) *The Independent*, 24 November 2002.

HARVESTING OF ILLEGAL DRUGS REMAINS HIGH (pages 98–99)

1. U.N. Office for Drug Control and Crime Prevention (ODCCP), *Global Illicit Drug Trends 2002* (New York: 2002), pp. 6–7, 47, 54–56; U.N. International Drug Control Programme (UNDCP), *Economic and Social Consequences of Drug Abuse and Illicit Trafficking*, UNDCP Technical Series, No. 6 (Vienna: January 1998), pp. 4–5.
2. UNDCP, op. cit. note 1, p. 5.
3. Synthetics growing rapidly from ODCCP, op. cit. note 1, p. 7; small share from UNDCP, op. cit. note 1, p. 5.
4. Peter Reuter and Victoria Greenfield, "Measuring Global Drug Markets: How Good Are the Numbers? Why Should We Care About Them?" *World Economics*, October–December 2001, pp. 159–73; "How Big Is The Drug Industry?" *NACLA Report on the Americas*, September/October 2002, pp. 10–11.
5. "Drugs, Guerrillas and Politicos in Mexico," *NACLA Report on the Americas*, September/October 2002, pp. 18–26; UNDCP, op. cit. note 1, p. 38..
6. UNDCP, "Cannabis as an Illicit Narcotic Crop: A Review of the Global Situation of Cannabis Consumption, Trafficking and Production," *Bulletin on Narcotics*, Issue 1, 1997, 1 December 1999.

7. ODCCP, op. cit. note 1, pp. 54–56.
8. Total for three nations from UNDCP, op. cit. note 1, p. 4; 75 percent from ODCCP, op. cit. note 1, pp. 54–56.
9. ODCCP, op. cit. note 1, pp. 54–56, 62.
10. Ibid., p. 55; Juan O. Tamayo, "Stepped Up Battle Against Coca Ignites Debate," *Miami Herald*, 16 April 2001; Kenneth E. Sharpe and William Spencer, "Refueling a Doomed War on Drugs, Flawed Policy Feeds Growing Conflict," *NACLA Report on the Americas*, November/December 2001, p. 25.
11. Scott Baldauf, "Afghans Try Opium-free Economy," *Christian Science Monitor*, 3 April 2001; ODCCP, op. cit. note 1, p. 16.
12. ODCCP, op. cit. note 1, p. 16.
13. Ibid., p. 6; Baldauf, op. cit. note 11; "The Poppies Bloom Again," *The Economist*, 20 April 2002.
14. ODCCP, op. cit. note 1, p. 6.
15. ODCCP, *Afghanistan Opium Survey 2002* (Vienna: October 2002), pp. 2–6; Tim Weiner, "With Taliban Gone, Opium Farmers Return to Their Only Cash Crop," *New York Times*, 26 November 2001; "The Poppies Bloom Again," op. cit. note 13; "Afghanistan Is Again the World's Largest Opium Producer, UN," *Associated Press*, 25 October 2002.
16. ODCCP, op. cit. note 1, pp. 45–51.
17. Ibid., pp. 6, 16.
18. UNDCP, op. cit. note 1, p. 1.
19. JoAnn Kawell, "Drug Economies of the Americas," *NACLA Report on the Americas*, September/October 2002, p. 8; UNDCP, op. cit. note 1, pp. 25–26.
20. "Profile: Colombia," *NACLA Report on the Americas*, September/October 2002, p. 13; "Profile: Mexico," *NACLA Report on the Americas*, September/October 2002, p. 14.
21. UNDCP, op. cit. note 1, pp. 25–26.
22. "Profile: United States," *NACLA Report on the Americas*, September/October 2002, p. 17; Jon Gettman and Paul Armentano, *1998 Marijuana Crop Report: An Evaluation of Marijuana Production, Value and Eradication Efforts in the United States* (Washington, DC: National Organization for the Reform of Marijuana Laws, October 1998.)
23. Matthew Brzezinski, "Re-engineering the Drug Business," *New York Times Magazine*, 23 June 2002; Reuter and Greenfield, op. cit. note 4, pp. 159–73; UNDCP, op. cit. note 1, pp. 12–14.
24. UNDCP, op. cit. note 1, pp. 12–14.
25. "Drug Economies of the Americas," *NACLA Report on the Americas*, September/October 2002, p. 11.
26. Brzezinski, op. cit. note 23.
27. "Afghanistan Is Again the World's Largest Opium Producer, UN," op. cit. note 15; ODCCP, op. cit. note 15, p. 40.
28. UNDCP, op. cit. note 1, p. 18.
29. Roberto Steiner and Alejandra Corchuelo, *Economic and Institutional Repercussions of the Drug Trade in Colombia* (Bogotá, Colombia: CEDE: Universidad de los Andes, December 1999), p. 14.
30. ODCCP, op. cit. note 1, p. 213. For this estimate, ODCCP defines drug abuser as a person who has used drugs at least once in the last year; idem, p. 282.
31. ODCCP, op. cit. note 1, p. 213.
32. UNDCP, op. cit. note 1, p. 9–10; ODCCP, op. cit. note 1, pp. 213–15.
33. UNDCP, op. cit. note 1, p. 8.
34. Ibid., pp. 8–11.
35. Ibid., pp. 36–37.

NUMBER OF REFUGEES DROPS (pages 102–03)

1. U.N. High Commissioner for Refugees (UNHCR), *Refugees by Numbers 2002*, at <www.unhcr.ch>, viewed 28 January 2003.
2. Ibid.
3. United Nations 1951 Convention Relating to the Status of Refugees, as amended by the 1967 Protocol Relating to the Status of Refugees, at <www.unhcr.ch/1951convention/timeless.html>.
4. UNHCR, op. cit. note 1.
5. Ibid.
6. Michael Flynn, "Searching For a Safe Haven," *Bulletin of the Atomic Scientists*, November/December 2002, pp. 24–25.
7. Marius Ergo, "Developing Countries Host Most Refugees," 15 November 2002, Norwegian Refugee Council, article on the UNHCR *Statistical Yearbook 2001*, at <www.nrc.no/pfweb/artikler/eng/UNHCR-151102.htm>, viewed 1 February 2003.
8. UNHCR, *Statistical Yearbook* 2001 (Geneva: 2002).
9. UNHCR, op. cit. note 1.
10. Ibid.
11. UNHCR, op. cit. note 8, p. 56; estimate based on an average of 1.4 persons per asylum case.
12. UNHCR, op. cit. note 8, p. 62.
13. Ibid., p. 46.
14. Ibid.
15. Ibid.

16 "Afghanistan, Agenda for Protection," Interview with Ruud Lubbers, U.N. High Commissioner for Refugees, *Refugees* (UNHCR), December 2002, p. 16.

17. Number of Palestinians from UNHCR, op. cit. note 1.

18. Global IDP Project, Norwegian Refugee Council, "Did You Know," *Forced Migration Review*, at <www.fmreview.org/2didyouknow.pdf>, viewed 1 February 2003.

19. Global IDP Project, *Internally Displaced People: Global Survey 2002* (London: Earthscan, 2002).

20. UNHCR, op. cit. note 1.

21. Ibid.

22. Ibid.

23. Ibid.

24. Ibid.

25. U.S. Committee for Refugees, "Principal Sources of Internally Displaced Persons," at <www.refugees.org/world/statistics/wrs01_table5.htm>, viewed 23 February 2003.

26. Environmental refugees are defined as "people who had to leave their habitat, temporarily or permanently because of a potential environmental hazard or disruption in their life supporting ecosystems"; Essam El-Hinnawi, in *Environmental Refugees* (Nairobi: U.N. Environment Programme, 1985), p. ii.

27. Norman Myers and Jennifer Kent, *Environmental Exodus: An Emergent Crisis in the Global Arena* (Washington, DC: Climate Institute, 1995).

28. Ibid.

29. Intergovernmental Panel on Climate Change cited in Alex Kirby, "West Warned on Climate Refugees," *BBC News Online*, 24 January 2000.

30. Ellen Read, Reuters, "Rising Seas Imperil Pacific Island Nations," *Environmental News Network*, at <www.enn.com/indepth/warming/impact7.asp>, viewed 12 January 2003.

31. Tim McGirk, "Environmental Refugees," *Time*, 31 January 2000.

32. Ibid.

33. World Commission on Dams, *Dams and Development: A New Framework for Decision-Making* (London: Earthscan, 2000), p. 104.

34. World Bank, *Involuntary Resettlement: The World Bank's Environmental and Social Safeguard Policies* (Washington, DC: March 2001).

35. UNHCR, op. cit. note 8.

36. U.S. Committee for Refugees, *World Refugee Survey 2002* (Washington, DC: 2002), p. 11.

37. Estimate of 50 percent from UNHCR, op. cit. note 8.; 80 percent estimate from Flynn, op. cit. note 6.

38. U.S. Committee for Refugees, *Funding Crisis in Refugee Assistance: Impact on Refugees* (Washington, DC: November 2002).

ALTERNATIVE MEDICINE GAINS POPULARITY (pages 104–05)

1. World Health Organization (WHO), *Report of the Interregional Workshop on Intellectual Property Rights in the Context of Traditional Medicine*, Bangkok, Thailand, 6–8 December 2000 (Geneva: 2001), p. 2; idem, *Traditional Medicine Strategy 2002–2005* (Geneva: 2002), p. 11.
2. WHO, *Traditional Medicine Strategy*, op. cit. note 1, pp. 9, 11.
3. Ibid.
4. Ibid., p. 2.
5. Ibid.
6. WHO, *Traditional Medicine Fact Sheet No. 134* (Geneva: September 1996).
7. WHO, *Traditional Medicine Strategy*, op. cit. note 1, p. 7.
8. Ibid., p. 1.
9. Ibid., p. 12.
10. Ibid., p. 13.
11. Ibid.
12. Ibid., p. 2.
13. Ibid., p. 11.
14. Ibid., pp. 11–12.
15. Carolyn Raffensperger, "Ecological Medicine," presentation at The Greening of Medicine Conference, San Rafael, CA, 17 October 2002.
16. Ibid.
17. Ibid.
18. Ibid.
19. WHO, *Traditional Medicine Strategy*, op. cit. note 1, p. 25.
20. Ibid.
21. Ibid., p. 22.
22. Ibid.
23. Ibid.
24. Ibid., p. 15.
25. Ibid., p. 9.
26. Ibid., p. 17.
27. Ibid., pp. 13, 32.
28. Ibid., p. 16.
29. Ibid., pp. 15–16.
30. Ibid., p. 18.
31. WHO, *Report of the Interregional Workshop*, op. cit. note 1, p. 6.

32. WHO, *Traditional Medicine Strategy*, op. cit. note 1, pp. 19–20.
33. Ibid.

MATERNAL DEATHS REFLECT INEQUITIES (pages 106–07)

1. World Health Organization (WHO), UNICEF, and U.N. Population Fund (UNFPA), *Maternal Mortality in 1995: Estimates Developed by WHO, UNICEF, UNFPA* (Geneva: 2001), p. 2.
2. UNICEF, *End Decade Databases—Maternal Mortality*, at <www.childinfo.org/eddb/mat_mortal/index.htm>, viewed 23 December 2002.
3. Elizabeth I. Ransom and Nancy V. Yinger, *Making Motherhood Safer* (Washington, DC: Population Reference Bureau, 2002), p. 6.
4. UNFPA, *State of World Population 2002* (New York: 2002), p. 35.
5. WHO et al., *Reduction of Maternal Mortality: A Joint WHO/UNICEF/UNFPA/World Bank Statement* (Geneva: WHO, 1999), p 13–14.
6. Ibid.
7. Ibid., pp. 16–17.
8. WHO, UNICEF, and UNFPA, op. cit. note 1, p. 2.
9. Ibid.
10. UNICEF, *End Decade Databases—Delivery Care*, at <www.childinfo.org/eddb/maternal/index.htm>, viewed 23 December 2002.
11. UNFPA, op. cit. note 4, p. 72.
12. World Bank cited in UNFPA, op. cit. note 4, p. 36. South Asia refers to India, Pakistan, Bangladesh, and Nepal.
13. World Bank cited in UNFPA, op. cit. note 4, p. 36.
14. Global Health Council, *Promises to Keep: The Toll of Unintended Pregnancies on Women's Lives in the Developing World* (Washington, DC: 2002), pp. 7–8.
15. Ibid.
16. Worldwatch calculation based on 700,000 maternal deaths over six years, from Global Health Council, op. cit. note 14.
17. Global Health Council, op. cit. note 14, p. 15.
18. Ibid.
19. Ibid., p. 41. Northern America refers to the United States and Canada.
20. UNICEF, op. cit. note 10.
21. Ibid.
22. Ibid.
23. Ransom and Yinger, op. cit. note 3, p. 9.
24. Ibid.

CONSUMPTION PATTERNS CONTRIBUTE TO MORTALITY (pages 108–09)

1. World Health Organization (WHO), *The World Health Report 2001* (Geneva: 2001), pp. 144–49. Mortality figures for 2001 show much the same pattern; data for 2000 are used here for comparison purposes, as WHO's detailed analysis by risk factor was done with data for the earlier year.
2. WHO, op. cit. note 1.
3. Ibid.
4. Ibid.
5. WHO, *The World Health Report 2002* (Geneva: 2002), p. 51.
6. WHO, op. cit. note 1, pp. 144–49.
7. WHO, op. cit. note 5, p. 57.
8. WHO, op. cit. note 1, pp. 144–49.
9. WHO, op. cit. note 5, pp. 224–27.
10. Ibid.
11. Ibid., pp. 69–70.
12. Ibid., p. 226.
13. Joint United Nations Programme on HIV/AIDS (UNAIDS) and WHO, *AIDS Epidemic Update* (Geneva: UNAIDS, 2002), p. 35.
14. WHO, op. cit. note 5, pp. 52–56.
15. Ibid., p. 68.
16. Ibid.
17. Global Alliance for Vaccines and Immunization, "Immunize Every Child: GAVI Strategy for Sustainable Immunization Services," at <www.gaviftf.org/forum/bb2/1-1_every_child.pdf>, viewed 31 January 2003.
18. Ibid.
19. Number without sanitation from Peter Gleick, "The Human Right to Water," *Water Policy*, vol. 1, no. 5 (1999), pp. 487–503; cost from WHO, op. cit. note 5, p. 128. WHO derived this cost using purchasing power parity, not market rates. Years of lost life refers to DALYs (disability-adjusted life years).
20. WHO, op. cit. note 5, p. 86.
21. Ibid.
22. Ibid., p. 226.
23. Judy Putnam et al., "Per Capita Food Supply Trends: Progress Toward Dietary Guidelines," *FoodReview*, September-December 2000, pp. 2–14.
24. WHO, op. cit. note 5, p. 226.
25. Ibid.
26. Ibid., p. 86.
27. Ibid.
28. Risk transition from ibid., pp. 4–6; Barry Popkin,

"An Overview on the Nutrition Transition and Its Health Implications: The Bellagio Meeting," *Public Health Nutrition*, February 2002, pp. 93–103.
29. WHO, op. cit. note 5, p. 86.
30. Soowon Kim, Soojae Moon, and Barry M. Popkin, "The Nutrition Transition in South Korea," *American Journal of Clinical Nutrition*, January 2000, pp. 44–53.
31. Puska Pekka et al., "Influencing Public Nutrition for Non-Communicable Disease Prevention: From Community Intervention to National Programme—Experiences from Finland," *Public Health Nutrition*, February 2002, pp. 245–51.
32. Ibid.

ORPHANS INCREASE DUE TO AIDS DEATHS (pages 110–11)

1. Joint United Nations Programme on HIV/AIDS (UNAIDS), UNICEF, and U.S. Agency for International Development (USAID), *Children on the Brink 2002: A Joint Report on Orphan Estimates and Program Strategies* (Washington, DC: TvT Associates, 2002), p. 5. This estimate is based on 88 countries for which data are available in Africa, Asia, and Latin America and the Caribbean.
2. Ibid., p. 6.
3. Ibid., p. 8.
4. Ibid., p. 5.
5. Ibid., p. 6.
6. Ibid., p. 5.
7. Ibid.
8. Ibid., p. 6.
9. Adult prevalence from UNAIDS, *Report on the Global HIV/AIDS Epidemic* (Geneva: July 2002), p. 190; orphans from UNAIDS, UNICEF, and USAID, op. cit. note 1, p.22.
10. UNAIDS, UNICEF, and USAID, op. cit. note 1, p. 22.
11. Ibid., pp. 7–8.
12. Ibid., p. 28.
13. Ibid., pp. 8, 22.
14. Ibid., p. 28.
15. Ibid., p. 6.
16. Ibid., pp. 22–23.
17. UNAIDS, *AIDS Epidemic Update: December 2002* (Geneva: 2002), pp. 7–9.
18. UNAIDS, UNICEF, and USAID, op. cit. note 1, p. 29.
19. Ibid., p. 6.
20. Ibid.
21. Ibid., pp. 24, 30.
22. Ibid., pp. 9–10.
23. Ibid.
24. Ibid., pp. 10–11.
25. Ibid., p. 12.

CORRUPTION THWARTS DEVELOPMENT (pages 114–15)

1. Transparency International, *Corruption Perceptions Index 2002* (Berlin: 2002).
2. Transparency International, "Frequently Asked Questions about the TI Corruption Perceptions Index (CPI) 2002," at <www.transparency.org/cpi/2002/cpi2002_faq.en.html>, viewed 3 February 2003.
3. Transparency International, *Global Corruption Report* (Berlin: 2003), p. 133.
4. Javier Corrales, "The Politics of Argentina's Meltdown," *World Policy Journal*, fall 2002, pp. 29–42.
5. Gallup International, "Governance and Democracy—The People's View: A Global Opinion Poll," *Gallup International Millennium Survey*, 1999, at <www.gallup-international.com/surveys1.htm>, viewed 7 January 2003.
6. Peter S. Green, "Graft in Eastern Europe is Called Rampant," *New York Times*, 7 November 2002; UNICEF survey of 40,000 children between the ages of 9 and 18 in 72 countries cited in UNICEF, *The State of the World's Children 2003* (New York: 2003).
7. Paulo Mauro, "Corruption and Growth," *Quarterly Journal of Economics*, vol. 110 (1995), pp. 681–713.
8. Shang-Jin Wei, *How Taxing is Corruption on International Investors?* Working Paper No. 6030 (Cambridge, MA: National Bureau of Economic Research, 1997).
9. World Bank, *Anticorruption in Transition: A Contribution to the Policy Debate* (Washington, DC: 2000), p. 19.
10. *Philippine Daily Inquirer*, 6 March 2002, cited in Transparency International, "Mugabe Stands Out Among the Politically Corrupt, While Banks and Energy Sector Top Dirty Business Deals Uncovered in 2002," press release (Berlin: 17 December 2002).
11. Transparency International, op. cit. note 3, p. 108, based on World Bank, *Report on Governance, Institutional Performance and Corruption: Developing an Anti-Corruption Strategy for Colombia*, 21 March 2002, and on <wobln0018.worldbank.org/LAC>.
12. George T. Abed and Sanjeev Gupta, eds., *Governance, Corruption, and Economic Performance* (Washington, DC: International Monetary Fund

(IMF), 2002).
13. Vito Tanzi and Hamid Davoodi, *Corruption, Public Investment and Growth*, Working Paper 97/139 (Washington, DC: IMF, 1997); Sanjeev Gupta, Hamid Davoodi, and Rosa Alonso-Terme, *Does Corruption Affect Income Inequality and Poverty?* Working Paper 98/76 (Washington, DC: IMF, 1998).
14. Transparency International, *Bribe Payers Index 2002* (Berlin: 2002).
15. Transparency International–Kenya, *Corruption in Kenya: Findings of an Urban Bribery Survey* (Nairobi: undated).
16. Yale University, "Environmental Sustainability Index," press release (New Haven, CT: 26 January 2001).
17. Charles Victor Barber et al., *The State of the Forest: Indonesia* (Washington, DC: World Resources Institute, 2002).
18. Neil Ford, "Oil: Ethics Vs. Profits," *African Business*, November 2000, pp. 26–27.
19. Floyd Norris, "A Nigerian Miracle," *New York Times*, 21 April 2002.
20. "Survey: Corruption, Construction, Conservatism," *Economist*, 20 April 2002.
21. *Asahi Shimbun*, 23 April 2002, cited in Transparency International, op. cit. note 3, p. 130.
22. Bernard Simon, "World Business Briefing Africa: Lesotho: Bribery Penalty," *New York Times*, 29 October 2002.
23. World Bank, "Listing of Ineligible Firms: Fraud and Corruption," at <www.worldbank.org/html/opr/procure/debarr.html>, viewed 7 January 2003.
24. John Brademas and Fritz Heimann, "Tackling International Corruption: No Longer Taboo," *Foreign Affairs*, September/October 1998, pp. 17–22.
25. Asian Development Bank, "Asian and Pacific Governments Adopt Regional Plan to Fight Corruption," press release (Tokyo: 30 November 2001).
26. Organization of American States, "Enhancement of Probity in the Hemisphere and Follow-Up on the Inter-American Program for Cooperation in the Fight Against Corruption," General Assembly Resolution, Washington, DC, 5 June 2000.
27. United Nations, General Assembly, "Revised Draft United Nations Convention Against Corruption," New York, 24 September 2002.
28. Demetrios Argyriades, "International Anticorruption Campaigns: Whose Ethics?" in Gerald E. Caiden et al., eds., *Where Corruption Lives* (Bloomfield, CT: Kumarian Press, 2001), pp. 217–26.
29. Transparency International, op. cit. note 14.
30. Peter Eigen, "Multinationals' Bribery Goes Unpunished," *International Herald Tribune*, 12 November 2002.
31. Ibid.

INTERNATIONAL CRIMINAL COURT STARTS WORK (pages 116–17)

1. Article 5, *Rome Statute of the International Criminal Court*, at <www.un.org/law/icc/index.html>, viewed 21 January 2003.
2. United Nations, Office of Legal Affairs, "Rome Statute of the International Criminal Court to Come into Force, Treaty Event to be Held at the UN Headquarters on 11 April," press release (New York: 2 April 2002).
3. Coalition for the International Criminal Court, *International Criminal Court: Basic Overview*, at <www.iccnow.org/documents.html>.
4. Roger S. Clark, *Countering Transnational and International Crime: Defining the Agenda*, Hume Papers on Public Policy (University of Edinburgh Press), vol. 6, nos. 1 and 2 (1998), pp. 20–29; Richard G. Wilkins, "The Right Thing The Wrong Way—Implications of the International Criminal Court," *Washington Times*, 1 October 2002.
5. Coalition for the International Criminal Court, *Questions and Answers on the International Criminal Court*, at <www.iccnow.org/documents.html>, updated 30 July 2002.
6. Leila Nadya Sadat and S. Richard Carden, "The New International Criminal Court: An Uneasy Revolution," *Georgetown Law Journal* , vol. 88 (2000), pp. 381, 384.
7. Gerard Prunier, *The Rwandan Crisis: History of a Genocide* (New York: Columbia University Press, rev. ed. 1997), pp. 261, 264–65.
8. John Kifner, "Inquiry Estimates Serb Drive Killed 10,000 in Kosovo," *New York Times*, 18 July 1999.
9. Barbara Crosette, "Military is Said to Prevent East Timor Refugees' Return," *New York Times*, 23 November 1999.
10. "The Pinochet Precedent: Who Could Be Arrested Next?" Interviews by Marguerite Feitlowitz, Crimes of War, at <www.crimesofwar.org/expert/pinochet-print.html>, viewed 21 January 2003.
11. Security Council Resolution 827, 1993 (establishing international tribunal for human rights violations in the former Yugoslavia); Security Council Resolution 955, 1994 (establishing international tribunal for

human rights violations in Rwanda).

12. "2003: The Year in Review," Crimes of War Project, at <www.crimesofwar.org/>, viewed 23 February 2003.
13. United Nations, "On the Establishment of the Special Court for Sierra Leone," at <www.un.org/Depts/dpko/unamsil/spcourt.htm>, viewed 23 February 2003.
14. David Cohen, *Seeking Justice on the Cheap: Is the East Timor Tribunal Really a Model for the Future?* AsiaPacific Issues Paper No. 61 (Honolulu, HI: East-West Center, August 2002).
15. "UN Resumes Talks on Cambodian Tribunal," *Associated Press*, 7 January 2003.
16. Pub. L. No. 101-513, 104 Stat. 1979 (1990); S.J. Res. 32, 103rd Cong. 139 Cong Rec. S930 (1993); War Crimes Act of 1995, H.R. 2587, 104th Cong. (1995).
17. See <www.un.org/icc/docs.htm>.
18. Steven Lee Myers, "U.S. Signs Treaty for World Court to Try Atrocities," *New York Times*, 1 January 2001.
19. Neil A. Lewis, "U.S. Rejects All Support for New Court on Atrocities," *New York Times*, 7 May 2002.
20. Ruth Wedgwood, "The Pitfalls of Global Justice," *New York Times*, 10 June 1998.
21. "Spain: Ratification of the International Criminal Court Rome Statute—One More Step Away from Impunity, One Step Closer to Justice," *Amnesty International News Service*, 26 October 2000.
22. Human Rights Watch, "The United States and the International Criminal Court," at <www.hrw.org/campaigns/icc/us.htm>, viewed 23 February 2003.

23 "2003: The Year in Review," op. cit. note 12.

24. "US Scores Win in India Against International Criminal Court," *Agence France-Presse*, 26 December 2002.

MILITARY EXPENDITURES ON THE RISE (pages 118–19)

1. Elisabeth Sköns et al., "Military Expenditure," in Stockholm International Peace Research Institute (SIPRI), *SIPRI Yearbook 2002: Armaments, Disarmament and International Security* (New York: Oxford University Press, 2002), p. 231. The estimate is based on adopted budgets; actual spending is expected to turn out higher because of subsequent supplemental budget requests. In general, official data often understate actual expenditures. Also, the number does not include data for nine countries for which no estimates are available.
2. Calculated from Sköns et al., op. cit. note 1, p. 231.
3. Ibid.
4. Ibid.
5. Ibid.
6. Ibid.
7. Ibid.
8. Ibid.
9. Ibid., pp. 231, 233, 235.
10. Ibid., p. 235.
11. Ibid.
12. Calculated from ibid., Table 6.A.3.
13. Ibid.
14. Ibid. SIPRI does not provide estimates for Cuba, Iraq, and Libya; data for these countries are from International Institute for Security Studies, *The Military Balance 2002–2003* (London: Oxford University Press, October 2002), Table 26.
15. Sköns et al., op. cit. note 1, pp. 254–55.
16. Center for Defense Information (CDI), "Fiscal Year 2004 Budget," at <www.cdi.org/budget/2004/top line.cfm>, viewed 3 February 2003.
17. Richard W. Stevenson and Elisabeth Bumiller, "President to Seek $48 Billion More for the Military," *New York Times*, 24 January 2002; James Dao, "Bush Sees Big Rise in Military Budget for Next 5 Years," *New York Times*, 2 February 2002.
18. CDI, op. cit. note 16; U.S. Department of Defense, Office of the Undersecretary of Defense (Comptroller), *National Defense Budget Estimates for FY2003* (Washington, DC: 2002). Figure 2 presents "budget authority" data, including nuclear weapons–related spending at the Department of Energy.
19. Leslie Wayne, "Rumsfeld Warns He Will Ask Congress for More Billions," *New York Times*, 6 February 2003; Jim Lobe, "Military Spending Inadequate, Hawks Tell Bush," *Inter Press Service News Agency*, 27 January 2003, at <www.ipsnews.net/print .asp?idnews=15464>.
20. CDI, op. cit. note 16.
21. Ibid.
22. White House, *The National Security Strategy of the United States of America* (Washington, DC: U.S. Government Printing Office, September 2002), p. 29.
23. Ibid., p. 15.
24. Elisabeth Bumiller, "White House Cuts Estimate of Cost of War with Iraq," *New York Times*, 31 December 2002.
25. William D. Nordhaus, "The Economic Consequences of a War with Iraq," in Carl Kaysen et al., *War*

With Iraq: Costs, Consequences, and Alternatives (Cambridge, MA: American Academy of Arts and Sciences, 2002), pp. 51–85.

26. Sköns et al., op. cit. note 1, pp. 233, 236–37.
27. Ibid., p. 237.
28. Calculated from Richard Grimmett, *Conventional Arms Transfers to Developing Nations* (Washington, DC: Congressional Research Service, various editions).
29. Sköns et al., op. cit. note 1, pp. 233, 236–37.
30. Ibid., p. 238.
31. Alex Bellos, "Brazil's New Leader Shelves Warplanes to Feed Hungry," *The Guardian*, 4 January 2003.
32. Ibid.
33. Dick Bell and Michael Renner, "A New Marshall Plan? Advancing Human Security and Controlling Terrorism," Worldwatch Institute, Washington, DC, 9 October 2001; World Game Institute, "Global Priorities," wall chart, at <www.worldgame.org>, 2002.
34. Dwight D. Eisenhower, "The Chance for Peace," address delivered before the American Society of Newspaper Editors, 16 April 1953, Dwight D. Eisenhower Library, at <www.eisenhower.utexas .edu/chance.htm>, viewed 11 February 2003.

RESOURCE WARS PLAGUE DEVELOPING WORLD (pages 120–21)

1. One-quarter share of all conflicts having a resource dimension is the author's assessment based on existing literature.
2. Number of deaths estimated from data in Milton Leitenberg, *Deaths in Wars and Conflicts Between 1945 and 2000* (College Park, MD: Center for International and Security Studies, University of Maryland, May 2001).
3. Some 350,000 deaths in the Congo are attributable to violence, the remainder to disease and malnutrition resulting from war disruptions; Taylor B. Seybolt, "Major Armed Conflicts," in Stockholm International Peace Research Institute (SIPRI), *SIPRI Yearbook 2002: Armaments, Disarmament and International Security* (New York: Oxford University Press, 2002), p. 33.
4. Refugee numbers derived from U.N. High Commissioner for Refugees, at <www.unhcr.ch>; number of internally displaced persons derived from U.S. Committee for Refugees, at <www.refugees.org>, both viewed 25 August 2002.
5. Michael Renner, *The Anatomy of Resource Wars*, Worldwatch Paper 162 (Washington, DC: Worldwatch Institute, October 2002).
6. U.N. Development Programme, *Human Development Report 2002* (New York: Oxford University Press, 2002).
7. Ibid.
8. Ibid.
9. Renner, op. cit. note 5.
10. Ibid.
11. Ibid.
12. Michael Renner, *Fighting for Survival: Environmental Decline, Social Conflicts, and the New Age of Insecurity* (New York: W.W. Norton & Company, 1996); Thomas F. Homer-Dixon, *Environment, Scarcity, and Violence* (Princeton, NJ: Princeton University Press, 1999); Thomas Homer-Dixon and Jessica Blitt, eds., *Ecoviolence: Links Among Environment, Population, and Security* (Lanham, MD: Rowman & Littlefield Publishers, 1998).
13. For specific cases, see Renner, op. cit. note 5.
14. Indra de Soysa, "The Resource Curse: Are Civil Wars Driven by Rapacity or Paucity?" in Mats Berdal and David M. Malone, eds., *Greed and Grievance: Economic Agendas in Civil Wars* (Boulder, CO: Lynne Rienner Publishers, 2000), pp. 120–21, 125–26; Jeffrey D. Sachs and Andrew M. Warner, "Natural Resource Abundance and Economic Growth," *Development Discussion Paper No. 517a* (Cambridge, MA: Harvard Institute for International Development, 1995); Michael Ross, *Extractive Sectors and the Poor* (Boston: Oxfam America, October 2001), pp. 5, 7–9.
15. William Reno, "Shadow States and the Political Economy of Civil Wars," in Berdal and Malone, op. cit. note 14, pp. 45–46, 56–57; de Soysa, op. cit. note 14, pp. 120–21, 125–26; Philippe LeBillon, "The Political Ecology of War: Natural Resources and Armed Conflicts," *Political Geography*, no. 20 (2001), pp. 561–84.
16. Ross, op. cit. note 14, pp. 13–14.
17. Small Arms Survey, *Small Arms Survey 2002* (New York: Oxford University Press, 2002); Michael Renner, *Small Arms, Big Impact: The Next Challenge of Disarmament*, Worldwatch Paper 137 (Washington, DC: Worldwatch Institute, October 1997).
18. Ongoing fighting from Marc Lacey, "War Is Still a Way of Life for Congo Rebels," *New York Times*, 21 November 2002; illegal networks from U.N. Security Council, "Final Report of the Panel of Experts on the Illegal Exploitation of Natural Resources and Other Forms of Wealth of the Democratic Republic of the Congo" (New York: 16 October 2002).

19. Renner, op. cit. note 5.
20. Ian Smillie, *Conflict Diamonds: Unfinished Business* (Ottawa, ON, Canada: International Development Research Centre, 27 May 2002); U.S. Government Accounting Office, *Critical Issues Remain in Deterring Conflict Diamond Trade* (Washington, DC: June 2002), pp. 17–21.
21. See, for example, the Campaign to Eliminate Conflict Diamonds, at <www.phrusa.org/campaigns/sierra_leone/conflict_diamonds.html>.

The Vital Signs Series

Some topics are included each year in *Vital Signs*; others are covered only in certain years. The following is a list of topics covered in *Vital Signs* thus far, with the year or years they appeared indicated in parentheses. Those marked with a bullet (♦) appeared in Part One, which includes time series of data on each topic.

AGRICULTURE AND FOOD

Agricultural Resources

♦ Fertilizer Use (1992–2001)
♦ Grain Area (1992–93, 1996–97, 1999–2000)
♦ Grain Yield (1994–95, 1998)
♦ Irrigation (1992, 1994, 1996–99, 2002)
Livestock (2001)
Organic Agriculture (1996, 2000)
Pesticide Control or Trade (1996, ♦2000, 2002)
Transgenic Crops (1999–2000)
Urban Agriculture (1997)

Food Trends

♦ Aquaculture (1994, 1996, 1998, 2002)
Biotech Crops (2001–02)
♦ Cocoa Production (2002)
♦ Coffee (2001)
♦ Fish (1992–2000)
♦ Grain Production (1992–2003)
♦ Grain Stocks (1992–99)
♦ Grain Used for Feed (1993, 1995–96)
♦ Meat (1992–2000, 2003)
♦ Milk (2001)
♦ Soybeans (1992–2001)
♦ Sugar and Sweetener Use (2002)

THE ECONOMY

Resource Economics

Agricultural Subsidies (2003)
♦ Aluminum (2001)
Arms and Grain Trade (1992)
Commodity Prices (2001)
Fossil Fuel Subsidies (1998)
♦ Gold (1994, 2000)
Illegal Drugs (2003)
Metals Exploration (1998, ♦2002)
♦ Metals Production (2002)
♦ Paper (1993, 1994, 1998–2000)
Paper Recycling (1994, 1998, 2000)
♦ Roundwood (1994, 1997, 1999, 2002)
Seafood Prices (1993)
♦ Steel (1993, 1996)
Steel Recycling (1992, 1995)
Subsidies for Environmental Harm (1997)
Wheat/Oil Exchange Rate (1992–93, 2001)

World Economy and Finance

- ♦ Agricultural Trade (2001)
- Aid for Sustainable Development (1997, 2002)
- ♦ Developing-Country Debt (1992–95, 1999–2003)
- Environmental Taxes (1996, 1998, 2000)
- Food Aid (1997)
- ♦ Global Economy (1992–2003)
- Microcredit (2001)
- ♦ Oil Spills (2002)
- Private Finance in Third World (1996, 1998)
- R&D Expenditures (1997)
- Socially Responsible Investing (2001)
- Stock Markets (2001)
- ♦ Trade (1993–96, 1998–2000, 2002)
- Transnational Corporations (1999–2000)
- ♦ U.N. Finances (1998–99, 2001)

Other Economic Topics

- ♦ Advertising (1993, 1999, 2003)
- Charitable Donations (2002)
- Cigarette Taxes (1993, 1995, 1998)
- Cruise Industry (2002)
- Ecolabeling (2002)
- Government Corruption (1999, 2003)
- Health Care Spending (2001)
- Pay Levels (2003)
- Pharmaceutical Industry (2001)
- PVC Plastic (2001)
- Satellite Monitoring (2000)
- ♦ Storm Damages (1996–2001, 2003)
- ♦ Television (1995)
- ♦ Tourism (2000, 2003)

ENERGY AND ATMOSPHERE

Atmosphere

- ♦ Carbon Emissions (1992, 1994–2002)
- ♦ Carbon and Temperature Combined (2003)
- ♦ CFC Production (1992–96, 1998, 2002)
- ♦ Global Temperature (1992–2002)

Fossil Fuels

- ♦ Carbon Use (1993)
- ♦ Coal (1993–96, 1998)
- ♦ Fossil Fuels Combined (1997, 1999–2003)
- ♦ Natural Gas (1992, 1994–96, 1998)
- ♦ Oil (1992–96, 1998)

Renewables, Efficiency, Other Sources

- ♦ Compact Fluorescent Lamps (1993–96, 1998–2000, 2002)
- ♦ Efficiency (1992, 2002)
- ♦ Geothermal Power (1993, 1997)
- ♦ Hydroelectric Power (1993, 1998)
- ♦ Nuclear Power (1992–2003)
- ♦ Solar Cells (1992–2002)
- ♦ Wind Power (1992–2003)

THE ENVIRONMENT

Animals

- Amphibians (1995, 2000)
- Aquatic Species (1996, 2002)
- Birds (1992, 1994, 2001, 2003)
- Marine Mammals (1993)
- Primates (1997)
- Vertebrates (1998)

Natural Resource Status

- Coral Reefs (1994, 2001)
- Farmland Quality (2002)
- Forests (1992, 1994–98, 2002)
- Groundwater Quality (2000)
- Ice Melting (2000)
- Ozone Layer (1997)
- Water Scarcity (1993, 2001–02)
- Water Tables (1995, 2000)
- Wetlands (2001)

Natural Resource Uses

- Biomass Energy (1999)
- Dams (1995)
- Ecosystem Conversion (1997)
- Energy Productivity (1994)
- Organic Waste Reuse (1998)

Soil Erosion (1992, 1995)
Tree Plantations (1998)

Pollution

Acid Rain (1998)
Algal Blooms (1999)
Forest Damage from Air Pollution (1993)
Hazardous Wastes (2002)
Lead in Gasoline (1995)
Nuclear Waste (1992, ♦1995)
Pesticide Resistance (♦1994, 1999)
♦ Sulfur and Nitrogen Emissions (1994–97)
Urban Air Pollution (1999)

Other Environmental Topics

Environmental Treaties (♦1995, 1996, 2000, 2002)
Nitrogen Fixation (1998)
Pollution Control Markets (1998)
Sea Level Rise (2003)
Semiconductor Impacts (2002)
Transboundary Parks (2002)
♦ World Heritage Sites (2003)

THE MILITARY

♦ Armed Forces (1997)
Arms Production (1997)
♦ Arms Trade (1994)
Landmines (1996, 2002)
♦ Military Expenditures (1992, 1998, 2003)
♦ Nuclear Arsenal (1992–96, 1999, 2001)
Peacekeeping Expenditures (1993, ♦1994–2003)
Resource Wars (2003)
♦ Wars (1995, 1998–2003)
Small Arms (1998–99)

SOCIETY AND HUMAN WELL-BEING

Health

♦ AIDS/HIV Incidence (1994–2003)
Alternative Medicine (2003)
Asthma (2002)
Breast and Prostate Cancer (1995)
♦ Child Mortality (1993)
♦ Cigarettes (1992–2001, 2003)
Drug Resistance (2001)
Endocrine Disrupters (2000)
Food Safety (2002)
Hunger (1995)
♦ Immunizations (1994)
♦ Infant Mortality (1992)
Infectious Diseases (1996)
Life Expectancy (1994, ♦1999)
Malaria (2001)
Malnutrition (1999)
Mental Health (2002)
Mortality Causes (2003)
Noncommunicable Diseases (1997)
Obesity (2001)
♦ Polio (1999)
Safe Water Access (1995)
Sanitation (1998)
Soda Consumption (2002)
Traffic Accidents (1994)
Tuberculosis (2000)

Reproduction and Women's Status

Family Planning Access (1992)
Female Education (1998)
Fertility Rates (1993)
Maternal Mortality (1992, 1997, 2003)
♦ Population Growth (1992–2003)
Sperm Count (1999)
Violence Against Women (1996, 2002)
Women in Politics (1995, 2000)

Social Inequities

Homelessness (1995)
Income Distribution (1992, 1995, 1997, 2002–03)
Language Extinction (1997, 2001)
Literacy (1993, 2001)
Prison Populations (2000)
Social Security (2001)
Teacher Supply (2002)
Unemployment (1999)

Other Social Topics

Aging Populations (1997)
Fast-Food Use (1999)
International Criminal Court (2003)
Nongovernmental Organizations (1999)
Orphans Due to AIDS Deaths (2003)
Refugees (♦1993–2000, 2001, 2003)
Religious Environmentalism (2001)
Urbanization (♦1995–96, ♦1998, ♦2000, 2002)
Voter Turnouts (1996, 2002)
Wind Energy Jobs (2000)

TRANSPORTATION AND COMMUNICATIONS

♦ Air Travel (1993, 1999)
♦ Automobiles (1992–2003)
♦ Bicycles (1992–2003)
Car-Sharing (2002)
Computer Production and Use (1995)
Gas Prices (2001)
Electric Cars (1997)
♦ Internet (1998–2000, 2002)
♦ Internet and Telephones Combined (2003)
♦ Motorbikes (1998)
♦ Railroads (2002)
♦ Satellites (1998–99)
♦ Telephones (1998–2000, 2002)
Urban Transportation (1999, 2001)

Worldwatch Publications

Signposts 2003

This new CD-ROM provides instant, searchable access to over 1,365 pages of full text from the last three editions of *State of the World* and *Vital Signs*, comprehensive data sets going back as far as 50 years, various historical timelines, and easy-to-understand graphs and tables. Fully indexed, *Signposts 2003* contains a powerful search engine for effortless search and retrieval. Plus, it is platform-independent and fully compatible with all current Windows (3.1 and up), Macintosh, and Unix/Linux operating systems.

State of the World 2003

Worldwatch's flagship annual is used by government officials, corporate planners, journalists, development specialists, professors, students, and concerned citizens in over 120 countries. In the 2003 edition, the Institute's highly respected inderdisciplinary research team shows how scaling up recent successes in curbing infection, increasing the income of the poor, and advancing the use of renewable energy, among others, would soon put the world's economy on a more sustainable path.

State of the World Library

Subscribe to the *State of the World Library* and join thousands of decisionmakers and concerned citizens who stay current on emerging environmental issues. The *State of the World Library* includes *State of World 2003* and *Vital Signs 2003* plus all four of the highly readable, up-to-date, and authoritative *Worldwatch Papers*.

Worldwatch Papers

Worldwatch Papers are written by the same award-winning team that produces *State of the World*. Each 50–70 page *Paper* provides cutting-edge analysis on an environmental topic that is making—or is about to make—headlines worldwide. Selected available *Papers* appear by topic below.

On Climate Change, Energy, and Materials

160: Reading the Weathervane: Climate Policy From Rio to Johannesburg, 2002
157: Hydrogen Futures: Toward a Sustainable Energy System, 2001
151: Micropower: The Next Electrical Era, 2000
149: Paper Cuts: Recovering the Paper Landscape, 1999
144: Mind Over Matter: Recasting the Role of Materials in Our Lives, 1998
138: Rising Sun, Gathering Winds: Policies To Stabilize the Climate and Strengthen Economies, 1997
130: Climate of Hope: New Strategies for Stabilizing the World's Atmosphere, 1996

On Ecological and Human Health

165: Winged Messengers: The Decline of Birds, 2003
153: Why Poison Ourselves: A Precautionary Approach to Synthetic Chemicals, 2000
148: Nature's Cornucopia: Our Stakes in Plant Diversity, 1999
145: Safeguarding the Health of Oceans, 1999
142: Rocking the Boat: Conserving Fisheries and Protecting Jobs, 1998
141: Losing Strands in the Web of Life: Vertebrate Declines and the Conservation of Biological Diversity, 1998
140: Taking a Stand: Cultivating a New Relationship With the World's Forests, 1998
129: Infecting Ourselves: How Environmental and Social Disruptions Trigger Disease, 1996

On Economics, Institutions, and Security

164: Invoking the Spirit: Religion and Spirituality in the Quest for a sustainable World, 2002
162: The Anatomy of Resource Wars, 2002
159: Traveling Light: New Paths for International Tourism, 2001
158: Unnatural Disasters, 2001
155: Still Waiting for the Jubilee: Pragmatic Solutions for the Third World Debt Crisis, 2001
152: Working for the Environment: A Growing Source of Jobs, 2000
146: Ending Violent Conflict, 1999
139: Investing in the Future: Harnessing Private Capital Flows for Environmentally Sustainable Development, 1998

On Food, Water, Population, and Urbanization

163: Home Grown: The Case for Local Food in a Global Market, 2002
161: Correcting Gender Myopia: Gender Equity, Women's Welfare, and the Environment, 2002
156: City Limits: Putting the Brakes on Sprawl, 2001
154: Deep Trouble: The Hidden Threat of Groundwater Pollution, 2000
150: Underfed and Overfed: The Global Epidemic of Malnutrition, 2000
147: Reinventing Cities for People and the Planet, 1999
136: The Agricultural Link: How Environmental Deterioration Could Disrupt Economic Progress, 1997

To order these and other Worldwatch publications, call us at 888-544-2303 or 570-320-2076, fax us at 570-320-2079, e-mail us at wwpub@worldwatch.org, or visit our Web site at www.worldwatch.org.

About the Worldwatch Institute

RESEARCH PROGRAMS

The Worldwatch Institute's interdisciplinary approach allows its team of researchers to explore emerging global issues from many perspectives, drawing on insights from ecology, economics, public health, sociology, and a range of other disciplines. The Institute's four research teams focus on:

- People
- Energy
- Nature
- Economy

PRESS INQUIRIES

Worldwatch provides reporters from around the world with access to the Institute's extensive research and the researchers behind it. For current information available to the media, visit our online press center at www.worldwatch.org/press.

For press inquiries or to be placed on the Worldwatch media list, contact Susan Finkelpearl by phone at 202-452-1992, ext. 517, by fax at 202-296-7365, or by e-mail at sfinkelpearl@worldwatch.org.

SPEAKERS BUREAU

Worldwatch researchers have extensive experience in bringing audiences up to date on important global trends, including food, water, pollution, climate, forests, oceans, energy, technology, and environmental security.

For more information, or to schedule a speaker, call Gary Gardner at 202-452-1992, ext. 521, or e-mail: ggardner@worldwatch.org.

INTERNATIONAL PUBLISHING PROGRAM

Worldwatch works with overseas publishers to translate, produce, and market its books, papers, and magazine. The Institute has more than 160 publishing contracts in over 20 languages. A complete listing can be found at www.worldwatch.org/foreign/index.html.

For more information, contact Elizabeth Nolan by phone at 202-452-1992, ext. 520, by fax at 202-296-7365, or by e-mail at enolan@worldwatch.org.

WORLDWATCH ONLINE

The Worldwatch Web site (www.worldwatch.org) provides immediate access to the Institute's publications. Save time and money by ordering and downloading Worldwatch publications in pdf format from our online bookstore. The site also includes press releases, special briefings on breaking environmental news, contact information, and job announcements.

SUBSCRIBE TO WORLDWATCH NEWS

Worldwatch maintains a free one-way e-mail list to distribute updates from the Institute as well as press releases on new books, papers, and magazine articles.

To subscribe, visit the Worldwatch Web site at www.worldwatch.org, or send an e-mail message to info@list.worldwatch.org and in the body of the message type: subscribe info.

FRIENDS OF WORLDWATCH

The Worldwatch Institute is a 501 (c)(3) non-profit organization. We rely on gifts from individuals and foundations to underwrite our efforts to provide the information and analysis needed to foster an environmentally sustainable society. Your gift will be used to help Worldwatch broaden its outreach programs to decisionmakers, build relationships with overseas environmental groups, and disseminate its vital information to as many people as possible through the Institute's Web site and publications.

To join our family of supporters, please call us at 202-452-1999 or send your tax-deductible donation to us at the address below. You can also donate online at www.worldwatch.org/donate.

LEGACY FOR SUSTAINABILITY

You can make a lasting contribution to a better future by remembering Worldwatch in your will. If you are interested in naming the Institute in your will, please contact us.

For further information on giving to Worldwatch, please contact Adrianne Greenlees by phone at 202-452-1992, ext. 518, by fax at 202-296-7365, or by e-mail at agreenlees@worldwatch.org.

Worldwatch Institute

VISION FOR A SUSTAINABLE WORLD

The WORLDWATCH INSTITUTE is an independent research organization that works for an environmentally sustainable and socially just society, in which the needs of all people are met without threatening the health of the natural environment or the well-being of future generations.

By providing compelling, accessible, and fact-based analysis of critical global issues, Worldwatch informs people around the world about the complex interactions between people, nature, and economies. Worldwatch focuses on the underlying causes of and practical solutions to the world's problems, in order to inspire people to demand new policies, investment patterns, and lifestyle choices.